Population Forecasting 1895–1945

European Studies of Population

VOLUME 5

The book series *European Studies of Population (ESPO)* aims at disseminating population and family research, with special relevance for Europe. It may analyse past, present and/or future trends,as well as their determinants and consequences.
The character of the series is multidisciplinary, including formal demographic analyses, as well as social, economic and/or historical population and family studies.
The following types of studies are of primary importance: (a) internationally relevant studies, (b) European comparative studies, (c) innovative theoretical and methodological studies, and (d) policy-relevant scientific studies. The series may include monographs, edited volumes and reference works.
The book series is published under the auspices of the *European Association for Population Studies (EAPS)*.

Editorial Offices:

Gijs *Beets*
Netherlands Interdisciplinary
Demographic Institute (NIDI)
P.O. Box 11650
NL – 2502 AR The Hague, Netherlands
Phone.: +31 70 356 5200
Fax.: + 31 70 364 7187
E-mail: beets@nidi.nl

Fred *Deven*
Population and Family
Study Centre (CBGS)
Markiesstraat 1
B – 1000 Brussels, Belgium
Phone: +32 2 507 3588
Fax: +32 2 507 3557
E-mail: cbgs@innet.be

Technical Editors:

Jacqueline *van der Helm*

Anita *Wouters*

The titles published in this series are listed at the end of this volume.

Population Forecasting 1895–1945

The Transition to Modernity

by

HENK A. DE GANS

University of Amsterdam,
Amsterdam Study Centre for the
Metropolitan Environment

KLUWER ACADEMIC PUBLISHERS
DORDRECHT / BOSTON / LONDON

A C.I.P. Catalogue record for this book is available from the Library of Congress.

ISBN 0-7923-5537-7

Published by Kluwer Academic Publishers,
P.O. Box 17, 3300 AA Dordrecht, The Netherlands.

Sold and distributed in North, Central and South America
by Kluwer Academic Publishers,
101 Philip Drive, Norwell, MA 02061, U.S.A.

In all other countries, sold and distributed
by Kluwer Academic Publishers,
P.O. Box 322, 3300 AH Dordrecht, The Netherlands.

Printed on acid-free paper

Special thanks to Stichting Dr Hendrik Muller's Vaderlandsch Fonds for generously supporting the publication of this book.

The *European Studies of Population book series* is published under the auspices of the **European Association for Population Studies** (EAPS), an organization which includes both individual and institutional members. Membership of the EAPS is open to anyone interested in or working in the area of European population studies, either in Europe or elsewhere. The Association organizes conferences, seminars and workshops, has initiated a number of working groups, and also publishes a semiannual newsletter. Members may subscribe to the *European Journal of Population* at a reduced rate.

EAPS Secretariat: EAPS, P.O. Box 11676, 2502 AR The Hague, The Netherlands;
tel. (+31-70) 356 52 39 / 356 52 29; fax (+31-70) 364 71 87.

Table of Contents

Foreword

Authors, scholars and scientists whose mother tongue is not one of the major languages of international communication are seriously disadvantaged. Some individuals, such as Joseph Conrad or Vladimir Nabokov, have overcome that handicap brilliantly. Others learn to live with it: they can express themselves sufficiently lucidly in a second language to make their voice heard internationally. At least when they have something original or striking to say they will be certain to reach their peers. Most scientists and scholars fall into that category. Others, again, have to wait until their work has been translated before its value is recognised. This may apply even to those whose mother tongue is widely read. The writings of Frenchmen Lyotard, Derrida, Baudrillard or Foucault on post-modernism, on language, discourse and power, for example, had tremendous world-wide impact only after English translations appeared on the market.

De Gans' study of the development of population forecasting in The Netherlands is another striking illustration of the effects a language barrier may have. He demonstrates convincingly that although a –possibly some what awkward– Dutchman named Wiebols, was a pioneer of modern cohort component demographic forecasting, he never received international recognition for this. In his thesis of 1925 Wiebols employed the newest instruments of demographic analysis in improving forecasting methodology. It was an important step in the transition from forecasts based on a mathematical extrapolation of some sort to forecasts based on the careful formulation of assumptions about trends in the components of change. Several of his fellow countrymen also made innovative contributions, but none of them found international recognition either: one will search in vain for their names in overviews of the development of the discipline. De Gans rightly stresses that this is not solely a consequence of the fact that they published in Dutch. Wiebols faced other obstacles too: his career in demography was frustrated, for example, when he was passed over for a job he wanted very badly. As a result he was unable to present his work internationally, and others failed to do so on his behalf. Even so, if his thesis had been in English his name would, no doubt, already have been listed with those of Cannan, Bowley or Wicksell.

It is the charm of De Gans' book that he does not simply recount the technical innovations made but that he presents these against a broad international backcloth and in the perspective of the demographic transition. He has an open eye for the developments in society which generated increased interest in the future course of population, as well as for the demands on the disciplines of demography, statistics and spatial planning resulting from that. And, while he may have saved some authors –both from the Netherlands and elsewhere in Europe– from oblivion, he does so by giving them a place in the Kuhnian paradigm shift in demographic forecasting which he sees unfold between 1895 and 1945.

In his years of research and reading Henk de Gans evidently developed sympathies and antipathies for his cast of characters. In a series of very readable vignettes he sketches the main protagonists and usually leaves one in no doubt about his assessment. He clearly does not want to share Wiebols' fate of having to depend on others to spread his fame as a scholar. By publishing his well researched study in English it will be accessible to the international demographic community. May it be widely read and consulted.

Dirk J. van de Kaa

Preface

Population forecasting 1895-1945: The transition to modernity was researched and written from 1990 onwards. Little did I realise in the early 1990s that my interest in the innovative performances of population forecasters of the inter-war period would bring me to the roots of population forecasting in early statistics and demography. Even less did I expect to find the 'prehistoric' endeavours in modern population forecasting of Edwin Cannan (1895) in England, Pontus Fahlbeck (1905) in Sweden and Harald Westergaard (1907) in Denmark, all evocative imaginations of the demographic transition in Europe, which merit a place in the intellectual history of demographic transition theory.

The modernisation of population forecasting resulted from the emergence and propagation of cohort component forecasting. Dutch forecasters of the inter-war period were pioneers in the modernisation of population forecasting methodology. Why their performances remained unnoticed in international histories is revealed in this book. One cause was the difficulty the Dutch representatives in the international field of statistics had in accepting probability calculus and the achievements of the mathematical school of statistics in population forecasting. One of the conclusions of my PhD thesis *Demographic Forecasting in the Netherlands 1895 1945: The Analysis and Implications of a Paradigm Shift,* published in 1997, was therefore that the difficulty these men had in accepting probability and uncertainty in calculating the future course of population demonstrated a Kuhnian paradigm shift in the study of future population.

I have received criticisms and suggestions for improvement from several quarters: the members of my doctoral committee, professors Roel den Dunnen, Andreas Faludi, and Anton Kuijsten of the University of Amsterdam, Henk ter Heide of the University of Utrecht and, in particular Nico Keilman of the University of Oslo; my colleague Harry van Vianen of the Population Research

Centre of the University of Groningen, and —not to be forgotten— the ESPO Editorial Board. Their thoughtful comments are highly appreciated. Publication in the European Studies of Population has offered the opportunity to reconsider the 1997 conclusion that the transition to modernity of 20th century population forecasting had the quality of a paradigm shift, by making use of these comments.

I am grateful to Dr Anne Hawkins of *Spels* in Utrecht who lent her skills in improving the linguistic accuracy and the style and readability of the English text. My special thanks and appreciation go to the Editorial Board of the European Studies of Population and Gijs Beets, Editorial Secretary of the series in particular, and to Tonny Nieuwstraten-Prins and Vanessa Gravekamp and all the other people of NIDI who have contributed to the production of this book.

Amsterdam, June 1998

List of vignettes

List of figures

List of tables

to political context. In some countries there was a close interaction between politics and science. In Fascist Italy, for instance, a symbiotic relationship existed from which both benefited (Ipsen, 1996, p. 4).

Reading G.A.H. Wiebols' thesis on the future size of the population of the Netherlands, the first of its kind in the Netherlands, is a fascinating experience (Wiebols, 1925). The searching quality of his work, the clarity of his explications, and the use of the newest instruments of demographic analysis are striking. Other Dutch pioneers of modern population forecasting, all dating from the inter-war period, contributed to the modernisation of population forecasting methodology. Interest in how inter-war assumption-making was influenced by social and economic conditions and curiosity of the backgrounds of the ongoing vehement debate on forecasting methodology in the Netherlands lay at the origin if this book.

The study focuses on the role played by Dutch forecasters in the history of the emergence and propagation of demographic forecasting methodology. The innovative performances of these men are set in the context of the transition of population forecasting to modernity, and against the international back cloth. The words modern and modernity need some clarification. In the preceding paragraph *modernity* was related to the increase of the statistics dependency of policy making and political decision making. But *modern* is also that which distinguishes the new from the old and anticipates the future. In the following chapters the word modern in modern population forecasting is understood in this sense. With population forecasting the total field of studies of population futures is meant, including for instance the mathematical extrapolation of total population. Demographic forecasting is used as a sub-category of population forecasting. It deals with both the calculation of future population based on the (demographic) components of population change (*cohort survival* and *cohort component* forecasting) and the future course of the development of the components themselves. Demographic forecasting and modern population forecasting are used as synonyms.[1]

[1] Contrary to current terminology *demographic forecasting* is used in this book as a sub-category of *population forecasting*. With *population forecasting* the total field of studies of population futures is indicated, conform to the general use in studies of futures, where *forecast* is used as a common denominator of all sorts of quantitative calculations of future population (Van Doorn and Van Vught, 1981). The term 'demographic forecasting' was not in use in the first four decades of this century, although Dutch urban and regional population forecasters started to speak of the *demographic method* in the inter-war period,

Contemporary efforts elsewhere reveal the comparative thoroughness of the quantitative aspects of Wiebols' demographic forecast, which served as the backbone of his study of the future course of the population of the Netherlands. Urban forecasts by Dutch town planners of the early 1930s were no less praiseworthy. The ingenuity of their elaboration is a convincing demonstration of the suitability of Wiebols' method for municipal forecasting and town planning purposes.

Dutch innovators of population forecasting in the inter-war period appear to have taken an international pioneering position. One would expect their achievements to be placed where they rightfully belong on the international scene. Surprisingly, contemporary overviews of literature and recent histories of population forecasting fail to refer to the achievements of the Dutch pioneers of demographic forecasting. How could this be?

1.2 | The Premise of a Dutch Pioneer Position in Population Forecasting

At first sight, on the basis of a study of histories of population forecasting (e.g. Wolfe, 1928/1929; Carr-Saunders, 1936; Glass, 1940) and contemporary overviews of forecasting literature (in France: Hecht, 1980; in Belgium: Wattelar, 1980, in Germany: Esenwein-Rothe, 1982 and Pflaumer, 1988) the premise of an international pioneer position finds no support. One would expect these written sources to abound with references to the Dutch innovative performances, but the opposite is the case. The sources are silent on this matter. Christine Wattelar's history of the modernisation of population forecasting, is typical:

---

in order to contrast it with economic forecasting. Demographic forecasting deals with both the calculation of future population based on the vital components of population change (to be compared with *cohort survival* and *cohort component forecasting*) and with the future course of the vital components themselves. The expression 'modern' population forecasting is not a very accurate synonym of demographic forecasting. Once demographic forecasting had become the new standard methodology in population forecasting, it was no longer modern in the sense of new. Calculations of future total population size based on *geometrical*, *logistic* or other mathematical growth functions are part of population forecasting, but not of demographic forecasting. In extrapolations calculated with mathematical growth functions arguments stem from a mathematical postulate, whilst demographic forecasts are based on knowledge of the interdependencies of structure and components of change (Esenwein-Rothe, 1982, p. 365).

From 1924 onwards, authors, Bowley in particular, broke with the habit of forecasting total population by global mathematical methods (geometrical forecasting) and to stress the importance of a fundamental element, the age structure of populations. In 1925, the Americans Dublin and Lotka demonstrated that the 'true' rate of population growth of the United States was below the observed growth rate, because of the impact of the influx of immigrants on the age structure of the population. Cannan had reasoned along similar lines in 1895. In 1926, the Swedish economist Wicksell was one of the first to introduce the component method for the prevision of the evolution of the population of Sweden, starting from assumptions regarding the components of population change separately. In 1928, the American Whelpton began a long series of previsions of the population of the United States. He was the first to start from a longitudinal point of view. He introduced the *cohort-survival method*, the cohort by cohort approach, and he applied the component method to regional population forecasts. In the same year, the French demographer Sauvy undertook his first attempt at demographic forecasting. His forecasts of 1932 and 1937 demonstrated the significance of such efforts, because he succeeded in making France aware of the risk of an immanent decrease in natality (Wattelar, 1980, p. 60).

One has difficulty in finding references to the presumed pioneer position in Dutch histories as well. With the exception of a concise overview by Van Praag (1977), demographic literature neglects the above mentioned inter-war pioneer endeavours. There are a few inventories with a short characterisation of employed methodology (e.g. CBS, 1951; Huijs, 1974). Until recent years there were no studies of a more searching nature.[2]

However, two Dutch sources appear to confirm the premise of an inter-war Dutch pioneer position in the innovation of population forecasting methodology explicitly. The town planner L.H.J. Angenot, an innovator of urban forecasting methodology in the 1930s himself, considered the domestic forecasts of the inter-war period to be of a much better quality than foreign forecasts. In discussing the process of demographic cohort-component forecasting, Godefroy, in his later years a professor of demography at the present Catholic University of Brabant in Tilburg, asserted that *'Wiebols was —as far as it can*

[2] Only Stolzenburg (1984) gives an extensive discussion of 20th century urban and regional forecasting as part of survey based planning. Arrears have recently been made good (e.g. De Gans, 1989, 1993, 1994a, 1995a, and 1995c; Ter Heide, 1992 and 1998).

be ascertained— the first in the world to do so' (Godefroy, 1960, p. 9). Like Angenot, Godefroy was an authority in population forecasting. The foundations of his commanding position were laid during the Second World War when he investigated contemporary international population forecasting methodology. To what extent the opinions of Angenot and Godefroy respectively were formed independently of each other remains to be seen. It is possible that Angenot based his opinion on views expressed by Godefroy, which he had based on an international literature review. On the other hand, it is not unlikely that Godefroy was influenced by Angenot's expertise in forecasting.[3]

Nevertheless, the corroborative utterances of Angenot and Godefroy support the premise of the pioneering position of Dutch population forecasters in the interwar period. These sources justify confidently proceeding in the search for explanations of the absence of references (below) and the assessment of the pioneering position by a comparative analysis of forecasts (Chapters 3 and 4).

Lack of knowledge of the Dutch contribution to the innovation of methodology can be attributed to several factors: Lack of interest among Dutch demographers and population forecasters; misinformation with respect to the Dutch

---

3 The adventures of the report of Godefroy's investigation bear witness of the obstacle that can occur in the propagation of knowledge. Godefroy worked as a secretary to the Committee for Regional Population Forecasts in the period 1943-1944. Angenot chaired the Committee. The Committee was set up by the Standing Committee of the National Physical Planning Agency in 1942 with the task of developing distribution methods for regional population forecasting purposes. Godefroy was asked to review the current international literature on population forecasting. While Secretary to the Committee, Godefroy also prepared his PhD thesis on the methods and techniques of population forecasting. When, however, his overview was finished, Godefroy refused to report to the Committee, much to the displeasure of the Chair. Godefroy perceived a conflict of interest between his thesis and his activity as Secretary of the Committee. Only on completing his thesis, in which the findings of the overview were used, was he ready to report his findings to the Committee. In April 1944, the Committee reluctantly decided to interrupt its activities for some months to give him the time and opportunity to complete his thesis. Presumably, because of the turmoil of war and its aftermath and Godefroy's personal involvement on the wrong side, the dissertation was never finished. When, in September 1946, the Committee recommenced its activities, Godefroy did not return as its secretary. Angenot's attempts to get hold of the report the Committee needed so much proved to be unsuccessful, too. The report was never published (Arch. RNP, inv. nr. 129, minutes of the meeting of December 13, 1943 and of April 18, 1944).

roots of modern population forecasting; one-sided orientation of Dutch authors to the international sources which failed to include information on Dutch pioneering endeavours. Lastly, to the rather one-sided orientation of (Dutch) population forecasting professionals on mathematical formulation, conceptualisation and modelling in the recent past decades. The orientation could have originated from the overvaluation of science, or –and more probably– from the specific interests of forecasting experts in the application of mathematics in forecasting (e.g. Yntema, 1977; Willekens, 1992).

The focus on the mathematical rigour of formulation in demography and population forecasting –that is, on formal (mathematical) theory making and modelling– has a long tradition, stemming both from Newton's idea of the description of nature in mathematical terms. The ultimate construction of a mathematical model of nature was considered to be the keystone in the arch of scientific knowledge (Casti, 1989, p. 21). To a high degree, the impetus of the modernisation of population forecasting came from mathematical statistics also and is discussed in Chapter 3. The mathematisation of demography and demographic forecasting gained momentum after World War II. The modelling part of population forecasting was increasingly taken over by mathematically trained experts interested in the mathematical rigour of forecasting models.[4]

An interesting example of this orientation is Willekens' paper on demographic forecasting commissioned by the Netherlands Organisation for Scientific Research. Keyfitz (1990) characterised Willekens' summary of where forecasting science stands and the directions in which it might move as masterly. Following the route of Wattelar and others, Willekens' pictures the course of development of demographic forecasting as follows: We owe the commonly used projection method to Leslie. Before him, Cannan (in 1895), Whelpton (in 1936) and others derived an analogous method, although with less mathematical rigour. The projection method has become known as the *component method*, a term invented by Whelpton. Leslie noted that the calculation model could be written as a system of simultaneous linear equations, presented

[4] Mathematisation manifested itself in the Netherlands in the 1930s, when the strife for rationalisation was directed at targets that could be realised: The economic plans of planning theoreticians; the techniques of quality control of industrial scientific managers; statistics by sociographers and demographers (Alberts, 1989).

compactly as a matrix multiplication. In this model, the population survives along cohort lines (Willekens, 1990a, p. 19).

Improvement of the mathematical rigour does not necessarily bring an improvement in methodology. In the next chapters it will become clear that, with few exceptions, Dutch forecasters of the inter-war period had little interest in a rigorous mathematical formulation of forecasting methodology, even though contemporaries thought highly of the mathematical aspects of their calculations. The focus was finding a sound quantitative basis for the discussion of population problems. Their calculations were difficult enough in the eyes of contemporary students of population with scant mathematical training. From a formal point of view, however, most forecasting models lacked rigour. This could explain the relative ignorance of the exploits of the non-formalising inter-war predecessors of post-war population forecasters.

There are several other explanations for the absence of references in international histories. The least convincing of these is the problem of language. The position of the Netherlands differs little from Scandinavia, and yet the names of Scandinavian contributions to the modernisation of population forecasting methodology are referred to abundantly. Apparently Dutch writers failed to propagate internationally information about the innovations of forecasting methodology in the inter-war period. Consequently, an answer must be found to the question What prevented the publication of the findings of the inter-war Dutch innovators of population forecasting methodology in international journals and at international conferences? This question pertains also to those, who were in the most favourable position to ensure the international propagation of knowledge of Dutch pioneering performances.

Presumably, specific character traits, views and private interests of the actors involved played a significant part in the absence of references. But such explanations do not suffice if, for instance, the French historians of demography, Dupâquier and Dupâquier (1985, pp. 393-394 and p. 416) are correct. They write in their *Histoire de la Démographie* that, from the end of the 19th century onwards, demographers and statisticians –the cadre to which most students of future population belonged– readily exchanged new ideas in an abundance of publications. According to the Dupâquiers, from the turn of the century onwards, innovations were known within months. The rapid propagation of new findings, theories and analytical instruments is one of the characteristics of modern science.

The student of modern Dutch population forecasting history is, however, confronted with a situation that contradicts the suggestion of the Dupâquiers. Easy propagation of the new methodology did not occur. If the premise that Dutch forecasters took a pioneering position is right, the absence of contemporary international references to their innovative endeavours indicates obstacles in the propagation of information about innovations. Explanations have to be in other factors.

1.3 | The Modernisation of Population Forecasting

The history of the substitution of one standard forecasting methodology by another describes how cohort component forecasting replaced mathematical calculations of the future size of population based on the simple assumption of a constant growth rate. The change was the result of a '*pursuit of objectivity*' in the national debate on the population problem in the 1920s (Van Praag, 1977, p. 259). As will become clear, the method of calculating future population size was much more than merely a technical matter. The issue lay at the heart of national and international debates on the population problem, in some countries as early as at the turn of the century and in most others, including the Netherlands, in the inter-war period. The rate of national population growth was central to the debates on the population issue. Calculations of future population growth and size formed the backbone of national and international discussions, kindling and, in a few instances, soothing deep-seated fears.

The substitution of the standard methodology resulted in a completely new approach to the calculation of future population development. From then on the focus was on the improvement of the understanding of the demographic foundations of population dynamics, and the impact of past and present population structure on future population growth.

If, notwithstanding the heat generated by the population issue, international scientific channels for the exchange of information were either unused, or obstructed, something important must have been going on. This book asserts that the causes explaining the absence of references to the Dutch pioneers of the new methodology can best be explained in terms of the occurrence of a paradigm shift in the study of future population as part of the modernisation of forecasting methodology (see section 1.4). The shift not only manifested itself in the emergence of a new forecasting method, known as demographic forecasting, but also in the obstacles to its acceptance, elaboration and propagation. The transition to modernity of population forecasting proceeded along the lines of the substitution of geometrical forecasting of total population size by demographic forecasting. The substitution process took about fifty years in all, from the first emergence of a demographic forecast in 1895 until about 1945. By that year, demographic forecasting had become the new standard forecasting methodology of national populations and, in the case of the Netherlands, was well on its way to become the standard method of urban and regional forecasting, too.

From a technical population dynamics point of view, the substitution of the standard methodology in question seems quite obvious. To the modern investigator, the substitution may even seem to be a process which had to happen, given the relative simplicity of the methods involved and the intrinsic analytical logic of the demographic approach. Surprisingly, however, this was not the case. The apparently simple replacement of a standard methodology proceeded along less obvious lines than might at first sight be expected.

Geometrical methodology offered an easy way to calculate the doubling time of a population at the present growth rate. Up until the 1920s, geometrical forecasting was popular among economists for this reason (Wolfe, 1928, pp. 537-542 and p. 678). The method was based on the assumption of the continuation of the observed crude growth rate of a population. The mathematical expression is $P_t = P_0 * (1 + r)^t$. At the turn of the century geometrical forecasting was characterised as the *standard* method (e.g. Shryock and Siegel, 1973; De Gans, 1994b; also Chapter 3). Until the 1930s, geometrical methodology and similar methods of calculating future total population with mathematical formulae found general application. Even nowadays the doubling time of populations is seen as a matter of interest (e.g. Lutz, Sanderson and Scherbow, 1997). Although it is still in use as a simple indicator of the rapidity of population increase, geometrical methodology has become obsolete as a forecasting instrument.

In the inter-war period geometrical forecasting methodology was replaced by demographic forecasting, the new standard by the end of the 1930s. In its most simple form, demographic forecasting was the extrapolation of observed time series of the vital components of population growth: The crude birth and death rates, and migration at mainly local and regional levels. In a more elaborate and modern form it was based on what was later to be called the *cohort-survival* and *cohort-component* approaches. These approaches involve the calculation of future population by age (and sex) categories. They start from a population by age or cohort (and sex) at a specific date and its components of change: Sets of age (or cohort) specific mortality and fertility rates.

In 1895, the English economist Edwin Cannan published a forecast of the population of England and Wales. In his forecast demographic forecasting in a rudimentary cohort survival form appeared for the first time. For that reason, the year 1895 is taken as the formal starting point of the era of demographic forecasting. Of course, as is explained in Chapter 3, even Cannan's forecast did not emerge out of the blue. Moreover, it took thirty years for demographic forecasting to re-emerge in England, or to be re-invented in the Netherlands (and other countries). It took a further fifteen years for demographic forecasting to be generally accepted and applied. The newness of the innovation of population forecasting methodology was long lasting. The Dutch statistician Methorst received national acclamation following the publication in 1937 of a study of the age pyramid as *'the pivot of the population problem'* in which he had emphasised the significance of age structure in understanding the dynamic foundations of the population problem (Methorst, 1937).

In the Netherlands, the substitution of standard forecasting methodology at the levels of the nation, the regions, and the large cities was completed by the year 1945. This year marks yet another transition. Prior to World War II Dutch national population forecasting had mainly been an activity of private persons interested in, or concerned about, the population problem. Population forecasting was cast into a government sanctioned organisational form during World War II, when the Government Service for the National Plan (the present National Physical Planning Agency) started its activities. Population forecasting took on public momentum when the Committee for the Regional Population Forecast (re)started its activities in 1946 (see also hapter 7). In the following years, Dutch government policy makers, influenced to a great extent by social democratic ideology, started to take planning seriously. Population forecasting became a necessary part of planning. The issue of regional population forecasting became a main field of interest of the National Physical Planning Agency. The Netherlands Central Bureau of Statistics (NCBS) became the authority responsible for national population forecasts. Due to the war its first forecast of the early 1940s, was never published. From 1951 onwards, official demographic forecasts have been made and published.

At urban and regional levels, the quest for estimating future migration and the consequent debates on the merits of economic forecasting versus demographic forecasting became the dominant issue in the 1930s, heralding a relatively brief era, from the second part of the 1940s up until the 1960s, of endeavours to forecast urban and regional population with socio-economic methodologies.

1.4 | A Paradigm Shift in the Study of Future Population?

A question to be addressed here is, why demographic forecasting took momentum only thirty years after its first appearance in 1895. The idea arose that the issue of the postponed re-emergence of demographic forecasting in the 1920s could satisfactorily be explained if it was assumed that factors had been at work that resisted an easy propagation of the new methodology. Moreover, if the assumption were true, similar factors could have been at work to prevent the international propagation of information of the methodological innovations made by Dutch forecasters. Possibly the assumption would afford a better understanding of the behaviour of certain key actors in the field of population forecasting and account for the above indicated absence in international history writing of the Dutch contribution to population forecasting. These considerations resulted in the presumption that the transition into modernity could even be characterised as a *paradigm shift* in the study of future population.

The paradigm shift concept, however, needs some clarification. The concept has a history of almost four decades in the sociology of science, starting with T.S. Kuhn's constructivist approach to science in his influential book on *The Structure of Scientific Revolutions* (1962/1970). Constructivist approaches have dominated the field of science and technology studies ever since. These approaches regard scientific facts as construed rather than discovered. Science should therefore be analysed as a social and cultural enterprise in order to improve our understanding of science. Constructivism became the common denominator for a variety of programs, theories and approaches in the decades following the publication of Kuhn's book (Hagendijk, 1996).

In Kuhn's conception of the development of science, *paradigm* stands for the whole set of fundamental ideas, principles, values and exemplary models shared by the members of a scientific community. Scientists carry out their day-to-day affairs within a framework of presuppositions about what consti-

Present day history of science has lost interest in such an idealised picture. In contrast, the focus of externalism is on the social and societal influences on the propagation of knowledge and takes the form of such questions as: How do specific conceptions become dominant? and How is consensus on specific claims of knowledge achieved? The externalistic approach results in critical intellectual histories, requiring the study of the development of population forecasting, for instance, to focus on the factors that affect and influence the emergence of new innovative methodology, its acceptance and propagation. Following Boudon, Keilman (1990, p. 55) makes a distinction between an empiricist and a nomological concept of knowledge. The dichotomy comes close to that of externalism versus internalism. The *empiricist* approach, to be compared with externalism, means the description of developments as they exactly have happened ("wie sie eigentlich geschehen sind"). The *nomological* type of knowledge, comparable with internalism, on the other hand involves the establishment of trends, regularities, cycles, laws governing which phenomena follow others, covariance between series, causal links between factors, et cetera.

There is a reserve among forecasters with respect to the past of population forecasting in general and population forecasts in particular. Forecasters are unfamiliar with looking back to past performances, for they deal with the future, not with past futures; they are driven to use the best information, assumptions, and techniques available. By having the state-of-the-art forecasting technology at their disposal, they are convinced of having accounted for most of the flaws of past assumptions. The most recent forecast is the forecaster's 'best shot', his former performances, are seen as having been overtaken by time. If any interest in the past of population forecasting exists at all, it is from an internalistic point of view: The sophistication of forecasting methodology in order to improve accuracy (Ascher, 1978, p. 8). An example of the internalistic approach is Keilman's investigation of uncertainty in population forecasting based on the analysis of the population forecasting endeavours of the NCBS since the 1950s (Keilman, 1990).

The efforts of population forecasters are mainly directed at improving the accuracy of their forecasts by focussing on better assumption making or the elaboration of forecasting models, although the impact of these efforts on accuracy has not been very impressive (Ascher, 1978; Keyfitz, 1981; Keilman, 1990). The internalistic approach is firmly directed at an evaluation and appraisal of forecasts and forecasting dynamics as forecasters themselves would do, that is, in terms of logical consistency, credibility of the assumptions and validity of the techniques. These evaluations often lead towards better specification and sophistication of the forecasting model, but not necessarily to a higher degree of accuracy of forecasts. The limited progress in the improvement of accuracy in population forecasting has led forecasters like Keilman and Ascher to suggest the broadening of the scope of the investigations in a more externalistic direction by taking into account the characteristics of the forecasters themselves.

Anchored Narratives

A means of overcoming some of the problems of the internalism-externalism dichotomy is the *anchored narratives approach*. It gives a forensic dimension to the recently emerged narratologic approach in history writing. The basic characteristic of the anchored narrative approach is the accentuation of the credibility of argument, to be compared with the relevant scientific community of Thomas Kuhn. A history can be told in many different ways, depending on the storyteller, the structuring principle that forms the backbone of his story, and the audience to which the story is addressed. Story, storyteller, structuring principle and audience are closely interrelated.

Historians and theorists of historiography became aware quite recently that not only the explications offered in a historical text, but also the text itself and the picture of the past that it presents can be subject of reflection. The awareness was awakened by the emergence of *narratology* in history in the 1970s. Point of departure in narratology is the conviction that a history is a historian's story of the past. Before the emergence of narratology, the language of a historical text was considered merely as a historian's obedient, willing instrument, serving the historian in reproducing the past. In narratology, the historical text –as well as the text of the historian– is as mysterious an object as the past itself (Ankersmit, 1990, p. 24).

A historian's story is not a mirror of historical reality, but the construction of a specific *coherence*, an interpretation of the past in which a specific coherence is construed. The coherence is not present in, or presented by, the past itself; it is the product of the story *(narratio)*. It is the historian who gives coherence to the past; the past is not his *discovery* (Lorenz, 1987, p. 100).[5]

---

[5] The views of Ankersmit and Lorentz resemble, among others, the dimension of the symbolically reconstructed past, which is part of G.H. Mead's philosophy of the present

The narratological approach to historical interpretation has been criticised on several points. It is argued that the past has a coherence of its own. Coming as close to it as possible is the historian's task. Narratology is criticised also for not answering the problem of the *truth* of a history where contradictory explanatory histories occur. Moreover, narratology only pertains to historiography and not to historical investigation (Lorenz, 1987, p. 100-106).

The focus on the forensic dimension of the *anchored narratives approach* can be taken as the historian's down-to-earth interpretation of the empiricist concept of knowledge, put in terms of the description of developments 'as they exactly have happened'. For the empiricist concept of knowledge leaves unanswered the question Who is to decide what exactly has happened. Recently, the anchored narratives approach has been introduced into demographic historiography (Van de Kaa, 1996). The forensic dimension pertains to narratives of the full story of what transpired, presented before a court of law. Evidently, in a forensic context a *good story* has to be plausible to the court and, taken in a wider context, to the audience. A story is plausible when it is complete, consistent, and anchored in evidence and common knowledge 'about the way things usually happen in this world' (Van de Kaa, 1996, p. 389). The demand for completeness in a well-anchored narrative may ask for elaboration in terms of sets of nested and hierarchically arranged sub-narratives that tell part of the full story in greater detail. The demand means that the evidence –the facts and the sources on which it is based– is accepted as such by the relevant audience, the court or the specific forum in question.[6]

The anchored narratives approach and Kuhnian theory have in common a constructivist character and a relativistic approach to the knowledge of reality. The problem of *truth* is not solved but it is accentuated that *truth* and *proof* are established in the interaction between the story teller and his audience.

---

and discussed in Chapter 8.

6 The metaphorical use of the forensic aspect of the concept can be stretched a little further. The forensic origins of the anchored narrative concept presuppose implicitly that the relevant audience (the court) must be convinced by the plausibility of the story that is told in the first place. In that case, not only the plausibility of the story, but also the characteristics such as the personality, rhetorical talents and societal position of the story teller play a part.

1.5 | The Elements of the Old and the New Paradigm

In the present study the change of standard forecasting methodology is taken as the structuring principle of the history of the transition to modernity of population forecasting in the 20th century. The classical standard of mathematical extrapolation of total population was substituted by demographic cohort-component forecasting. In the course of this book it will be made clear that there are sufficient indications that the change of standard method involved more than a mere change of method. It entailed a change of the perception of the workings of Nature in population growth. For that reason the classical and the new forecasting methodology are considered to be part of two different paradigms in the study of the future of population. The main elements of these paradigms, summarised in *Table 1.1*, are discussed here.

Basically, the old standard forecasting methodology rested on the belief that population growth is governed by a 'law' of total population growth. The law is expressed in a mathematical growth model (a geometric growth function; in the 1920s a logistic growth function).[7] Migration is included in the crude growth rate, but is not distinguished separately. In the wake of the Malthusian belief in a geometrical law of population growth, most economists of the 19th century were interested in the results of population increase rather than in its mode. No major efforts were made to inquire whether the growth rate was constant or not (Wolfe, 1928, p. 678). By implication it can be assumed that the belief in the existence of a geometrical law must have been feeble.

The forecasting method that substituted the classical standard rested on a different belief, namely that the future course of population is a matter open to speculation. The new method necessitated theories of population in order to ascertain the direction in which population is ultimately heading. For the short and middle term, the (demographic) characteristics of population dynamics and specifically the impact of past and present age structure on the future numbers of births and deaths, and consequently on future population growth had to be understood. Gradually, the calculation model developed into what is known today as the *cohort survival* and *cohort component* models (in formal terminology, Leslie's matrix model). The basic difference between the

---

7 The mathematical expression of the logistic growth of population is $P(t) = K/(1 + Be^{-rt})$ (Rogers, 1985, p. 20). The history of the logistic growth concept goes back to the Belgian mathematician François-Pierre Verhulst in 1840.

classical and the new approach to forecasting can be characterised with the words 'law' versus 'speculation' or 'law' versus 'theory of population'.

The concept of law needs some explanation. As it was put by Bouthoul –almost at the end of the period of the transition to modernity of population forecasting– forecasting (Fr.: 'prévision') and scientific laws are conflicting concepts (Bouthoul, 1935). There are two interpretations of the concept of scientific law, resulting in different conceptions of forecasting. In its most simple form, a scientific law is seen as a concept of a uniform development taking place according to a pre-existing plan, as in the Newtonian philosophy of life. The acceptance of this interpretation implies that the future is laid down beforehand. Forecasting is limited to understanding what already exists, but is still hidden from our eyes. In this case forecasting is nothing but discovering what has already been decided or determined.

If this interpretation is followed, there is no great difference between explanation, prediction and testing. As Popper puts it: "The difference is not one of a logical structure, but rather one of emphasis; it depends on what we take it to be our problem and what we do not so consider. If it is not our problem to find a prognosis, while we take it to be our problem to find the initial conditions or some of the universal laws (or both) from which we may deduce a given 'prognosis' then we are looking for an explanation (and the given 'prognosis' becomes our 'explicandum').If we consider the laws and initial conditions as given (rather than as to be found) and use them merely for deducing the prognosis, in order to get thereby some new information, then we are trying to make a prediction." (Popper, 1960, p. 133; also, for a discussion of forecasting from a epistemological perspective, Van Seventer and Vossen, 1979).

This interpretation of law is at the basis of the classical standard approach to forecasting future population and plays an important part in the chapters that follow. The interpretation implies the belief that the future course of total population can be predicted with certainty.

In a more sophisticated interpretation, however, a scientific law is seen as a *relational* concept. This view rests on the assumption of the existence of a relationship between two different orders of facts. The relationship is neither arbitrary nor accidental and has a more or less permanent character. The encounter, the combination of two orders of facts, brings about a specific reac-

Table 1.1. The elements of the old and the new paradigm in the study of future population

Elements	Paradigm (old)	Paradigm (new)
Scientific community	actuaries economists *statisticians* town planners	actuaries economists *statisticians* town planners
Central belief with respect to the future course of population	determined by a 'law' of population growth	*a matter of speculation *determined by population dynamics (short and middle term) *dependent of theory of population (long term)
Perception of reliability of the results	certainty	uncertainty
Character of the future	closed	open
Interest in future population directed at	*total population size *crude growth rate	*age-sex structure of population *age-sex specific occurrence/exposure rates of components of change *total population size
Migration	included in total growth rate	omitted (in most instances)
Forecasting model	mathematical *geometrical *logistic (in 1920s)	demographic *cohort-survival model *cohort-component (= Leslie-model)
Task forecaster directed at finding	*the mathematical expression of the law of population growth *region specific parameters of the mathematical equation	*a theory of population constructing a model of population dynamics *determination of probable direction of the future course of the vital rates
Perception of position of forecaster in forecasting process	*influence personal judgement limited *external to the population system that is forecast	*influence of his personal judgement is considerable *he is part of the system that is forecast

tion that can be known, specified and predicted. With respect to forecasting, such a law means the acceptance that order exists, but also that no unique, unavoidable direction follows from the course of development of the phenomena involved. A change of direction results from changes of the interplay of the main influencing elements, either by the occurrence of new elements, or by the repression of existing elements, or by variations of numbers, intensities, frequencies, and so forth. Consequently, to retain its scientific character, a forecast cannot escape making reservations the moment it ventures into the domain of the future, even if it were to succeed in making a complete inventory of all the relevant factors, facts and influences pertaining to a phenomenon in a specific field. Such a reservation is always of the 'ceteris paribus' type (Bouthoul, 1935, pp. 222-223).

In Bouthoul's second interpretation 'law' is part of the sequence of the logical structure of the scientific method: Observations/facts > laws > theories > models. Laws show a functional relationship between two or more kinds of events, but do not tell why the relationship exists. For this we need a theory. The second interpretation brings us in the domain of the new approach to forecasting future population. The focus is in the first place on understanding the mechanism of population dynamics, namely the contribution of the separate components of population growth to total population growth; the interaction of population composition (the age-sex structure of a population) and the respective growth components (mortality, fertility, migration). Secondly, on understanding the interaction between social-economic processes and demographic processes. Lastly, when long-term forecasting comes into play, forecasters could not do without a theory of future population. This means that they had to have an idea (an expectation, a belief or a theory) about the direction a population was ultimately heading for.

Certainty and confidence (the main characteristics implied in the classical methodology) were substituted for understanding population dynamics and speculation (theory) about the future course of population.

As is shown in the next chapters, the scientific community of students of (future) population is diffuse. The community consists of scientists interested in the study of population and familiar with mathematics. They work as actuaries, official and mathematical statisticians, economists or (town) planners. A central position is taken by official statisticians, such as the chiefs of statistical offices. In the Netherlands the most outstanding representatives of this category were jurists, who lacked a sound mathematical training.

There was no separate population forecasting profession. Notwithstanding, some kind of *institutionalisation* has taken place in the inter-war period. An intellectual activity is institutionalised, if there is a relatively dense interaction of persons who perform that activity. A high degree of institutionalisation entails its teaching and investigation within a regulated, scheduled, and systematically administered organisation. It also entails the organised support of the activity from outside the particular institution and the reception or use of the activity beyond the boundaries of the institution (Shils, in Hodgson, 1991, p.21 n. 21). In Chapters 4 and 5 it will be shown that there was a relatively dense interaction of persons involved in demographic forecasting, particularly in the 1930s; there was an exchange of information about new developments in publications and at conferences; scientific debates on central issues were organised; the nature, the societal position and the main tasks of the intellectual activity of population forecasting were reflected upon; and, lastly, forecasting methodology was eventually discussed in textbooks for students of economy and statistics.

1.6 | The Composition of this Book

Initially, the idea was to write a history of demographic forecasting in the Netherlands in the first four decades of the 20th century, and highlighting the contributions of the innovators of the field. It soon became clear that the history coincided with the history of the emergence and propagation of demographic forecasting. The project started from the premise that the Dutch pioneers of demographic forecasting were international pioneers as well. Because no traces of such an international pioneering position could be found, neither in international contemporary histories nor in more recent ones, a new dimension was added: Finding an explanation for the absence of international references to these innovative contributions. Consequently, the history took on the form of a more general history of the transition to modernity of population forecasting in the first half of the 20th century. It soon became clear that hindrances must have been at work, that prevented an easy propagation of knowledge of innovations of forecasting methodology. Because of these hindrances the suggestion was raised that processes of a fundamental nature, that could be reduced to the common denominator of a change of paradigm in the study of future population, had been at work. It was even thought, that the presumed change of paradigm might have taken the proportion of a paradigm shift, in the sense of Kuhn's theory of scientific revolutions, giving the present history a science dynamics dimension.

The composition of the book is not strictly chronological but combined chronological and topical. There is chronology within the chapters, but not necessarily between them. Each chapter is a narrative in itself, written from a specific perspective. After the present introduction (Chapter 1), the most important pioneer of demographic forecasting in the Netherlands, G.A.H. Wiebols, is introduced; his failure to achieve a professional position and an international reputation is explained, and the consequences are set out (Chapter 2). Next, the focus is on the 'prehistory' of demographic forecasting: The conceptualisation of population dynamics; its roots in statistics and the first endeavours in demographic forecasting (Chapter 3). The true breakthrough of demographic forecasting is in the inter-war period. The period is the setting of the conflict between the old paradigm –in the guise of a renaissance of the belief in a law of population growth– and the new paradigm, in their pursuit for dominance (Chapter 4). The modernisation of population forecasting in the Netherlands took place in the 1920s when Wiebols and others laid the foundations of the demographic approach. The content, context and characteristics of the debate on demographic forecasting methodology are analysed and discussed in Chapter 5. However, the advancement of the methodological innovation of demographic forecasting came from housing and town planning in the course of the 1930s. First, the advancement of demographic forecasting methodology at urban and regional levels within the context of housing is discussed (Chapter 6). Impulses for the application and innovation of forecasting methodology came also from the demands of town extension plans that had to provide for a forecast number of persons, households or dwellings. It became clear that urban demographic forecasting could easily and directly be applied in various areas of policy making as, for instance, regional planning, urban development planning and housing policy (discussed in Chapter 7). The change of paradigm in population forecasting had far-reaching implications for the conception of time in forecasting and consequently for the position of the forecaster in the forecasting process as well as in society (Chapter 8). Finally, the main findings are summarised and the question whether the transition of population forecasting to modernity was a matter of a change of paradigm or of a paradigm shift in terms of Kuhnian theory is answered and the implications of the main findings for present and future population forecasting reconnoitred (Chapter 9).

2. A Dutch Pioneer of Demographic Forecasting:

The Story of G.A.H. Wiebols (1895-1960)

2.1 | An Answer to a Promising Invitation from Amsterdam

On 10 June 1925, the young doctor Wiebols, who just four weeks before had publicly defended his PhD thesis on the future size of the population of the Netherlands at the Business School of Rotterdam (at present Erasmus University), received a letter from the distinguished Dutch statistician, Dr. J.H. van Zanten, Director of the Amsterdam Bureau of Statistics (ABvS). That letter has been lost, but we can deduce from Wiebols' responses of 17 June and 2 July, presented below, that Van Zanten had complimented him on his dissertation, while taking the opportunity to react to Wiebols' criticisms of the quality of Dutch public statistics. Van Zanten invited Wiebols to specify what was lacking in the public statistics of the municipality of Amsterdam to facilitate forecasting of Amsterdam's future population in line with the methodological directions given in his thesis. Wiebols had indicated the statistical conditions to be fulfilled for a further specification of a national forecast (Wiebols, 1925, pp. 110-127). His excursions into the possibilities of further elaboration of his method prompted Van Zanten to invite him to explore the forecasting possibilities of the public statistics of Amsterdam.

Wiebols was clearly delighted with Van Zanten's letter. It is interesting to note how he wished to allay any impression that he had intended to criticise the quality of the statistics published by Van Zanten's Amsterdam Bureau of Statistics.[8] The letters demonstrate also how ready Wiebols was to accept the invitation to indicate what statistical conditions had to be fulfilled for a more

[8] This part of his letter is not inserted in the abstracts of Wiebols' letters.

specific population forecast of the population of Amsterdam. Van Zanten's reaction was a recognition by an authority in the field of statistics and demography of Wiebols' performance; without doubt, it ought to have transformed his future employment prospects. He had experienced difficulty in finding a job on graduating from Rotterdam university and had been unemployed for several years. He had made the best of this period by working on his dissertation.[9]

2.2 | Wiebols the Innovator

What was so special about Wiebols' work? At this stage we merely reveal that he made the first elaborate demographic forecast of the population of the Netherlands, based on the age-sex structure of the female part of the population, age-sex specific survival rates derived from dynamic life table probabilities, dynamic general fertility rates and a thorough foundation of the assumptions of the future behaviour of these age-specific components of population growth. Moreover, he was the first in the Netherlands, if not in the history of modern population forecasting, to use net reproduction in population forecasting, several years before Kuczinski got international renown with the introduction of the Net Reproduction Rate (NRR). Wiebols had indicated that the actual net reproduction of the Dutch population was about 46 per cent above the level of mere replacement of the population, and that the Netherlands were facing considerable population growth. These insights were used to argue that working with age-specific fertility rates was absolutely necessary in population forecasting (Wiebols, 1925, pp. 41-42). His letters to Van Zanten give a concise impression of his approach.[10]
On the base of his calculations he arrived at a population total of about 9.5 million inhabitants in 1950 and 12.7 million in the year 2000, compared with

---

9 Gerhard Adolf Heinrich Wiebols was born in Vlaardingen in 1895. He took his school leaving examinations in 1913. Shortly afterwards he served in the army (from 1914-1917), because of the mobilisation of the Netherlands during the First World War. On leaving the army he studied commercial science at the Business school of Rotterdam. He graduated in 1921. After his graduation Wiebols had difficulty in finding a regular occupation. He left the Netherlands for Germany, where he stayed for one year. Back in Holland and unemployed, he worked on his PhD thesis on the future population size in the Netherlands. His book was published in 1925.

10 A discussion of Wiebols' contribution from an international perspective is given in Chapter 4. In Chapter 5 his performance is placed in the perspective of the history of national population forecasting in the Netherlands. On the intellectual history of NRR, see Vignette 5.

Wiebols' letter to Van Zanten[11]

Dear Sir, (June 17, 1925)
Many thanks for your kind letter of the tenth of June. I am sorry that it took some time before I could write an answer, because I was absent for a few days. [....]

The purpose of my thesis was to give an estimate of the future size of the population of the Netherlands. I started from the assumption that the past development would continue in the future and asymptotically approximate to an assumed minimum. In order to determine the minimum of fertility, I related to abroad (in casu, France); with regard to mortality to (the life table of, HAdG) public servants, a population category living under very favourable conditions of mortality.

But otherwise, only statistical data of the national population total were available for my problem. As long as public statistics of the Netherlands as a whole do not allow for further specification, my problem is not served if a few other statistics do allow such a thing. Consequently, further research into this matter was superfluous. Even if, for instance, statistics of Amsterdam allow for specification, this would not hold for the Netherlands too; not so much because the numbers would become too small (in principle, this problem could be circumvented by taking into account larger periods instead of ten-year periods), but because the population of Amsterdam is a very specific population category, differing as much from the population of the Netherlands as any foreign population would do. [...]

Dr. G.A.H. Wiebols

Consequently, it appears to me that the public statistics of Amsterdam, even if they allow for more specification than those of the Netherlands, cannot directly be exploited for the problem that I have posed myself: The estimation of the future size of the population of the Netherlands. However, they can be exploited for a problem that is analogous to the previous

11 Abstracts from Dr. G.A.H. Wiebols' letters to Dr. J.H. van Zanten, director of the Amsterdam Statistical Office dating from June 17 and July 2, 1925 (GAA, Arch. 5185/1925/nr.366).
Photograph Dr. Wiebols from album Mr. Paul Wiebols (Zierikzee, Netherlands).
Letter and diagram unauthorized translated by Henk de Gans.

one, the estimation of the future size of the population of Amsterdam. It is of the utmost scientific importance that such an estimate be made. Apart from its significance for the population problem of Amsterdam, that of the Netherlands as a whole would profit as well, albeit indirectly. For if one has an estimate of the future sizes of both the population of the Netherlands and Amsterdam (more generally: The big cities) and if in both estimates the same method is followed and the same delimitation of boundaries is kept in mind, then the future size of the population of the country could be deduced. If the outcome would appear to be unlikely from the present point of view, then it would follow that the estimated future size of the population of the Netherlands is unlikely too. For if the public statistics of Amsterdam (and those of the big cities in general) allow for a further specification than those of the Netherlands, the outcome deserves more confidence than an outcome based on statistics of the Netherlands. If the two outcomes are not in harmony with each other, the latter (those of the Netherlands- HAdG) should be revised and rectified.

An estimate of the future size of the population of Amsterdam raises difficulties that do not arise in the case of an estimate of the national population, because:

1. as is already mentioned in your letter, numbers may become too small, obliging to take longer projection intervals. Apart from the difficulty of averages becoming less meaningful, there is also the technical difficulty that when larger intervals are taken, developments have to be pursued further back into the past. Developments, that might be detected if smaller time intervals were considered, may then get lost;
2. migration –which can be neglected for the Netherlands because of its insignificance– cannot be excluded as a factor because of its importance in Amsterdam. As a consequence a very speculative element is introduced in the estimation of the future size of the population of Amsterdam.

However, I do believe that the difficulties are amply outweighed by the advantages of a further elaboration and that you would render science a great service if you would provide the occasion for that purpose.

This brings me to consider a second question. It indicates what in my opinion is lacking in Amsterdam statistics if a further specification is envisaged, and before long I hope to inform you of my conclusion. Therefore, I request you kindly to grant me some time, since my duties trespass on my evenings too, because of a rush of business of drawing up the balance sheet, leaving me not much time for scientific work.

(July 2, 1925). As far as is possible, I am now able to answer the second question of your letter of June. I have interpreted the question as follows: To examine what is wanting in the public statistics of Amsterdam for allowing an estimation of the future size of the population of Amsterdam that rests on a specification that goes further than my estimation of the future size of the national population.

Such a specification is given in the annexed diagram. All factors mentioned relate to the future, unless it is explicitly mentioned that the factors hold for the present situation (December 31, 1920). All factors are specified by age groups. Of course, other specifications are possible too, but in my opinion the one given here

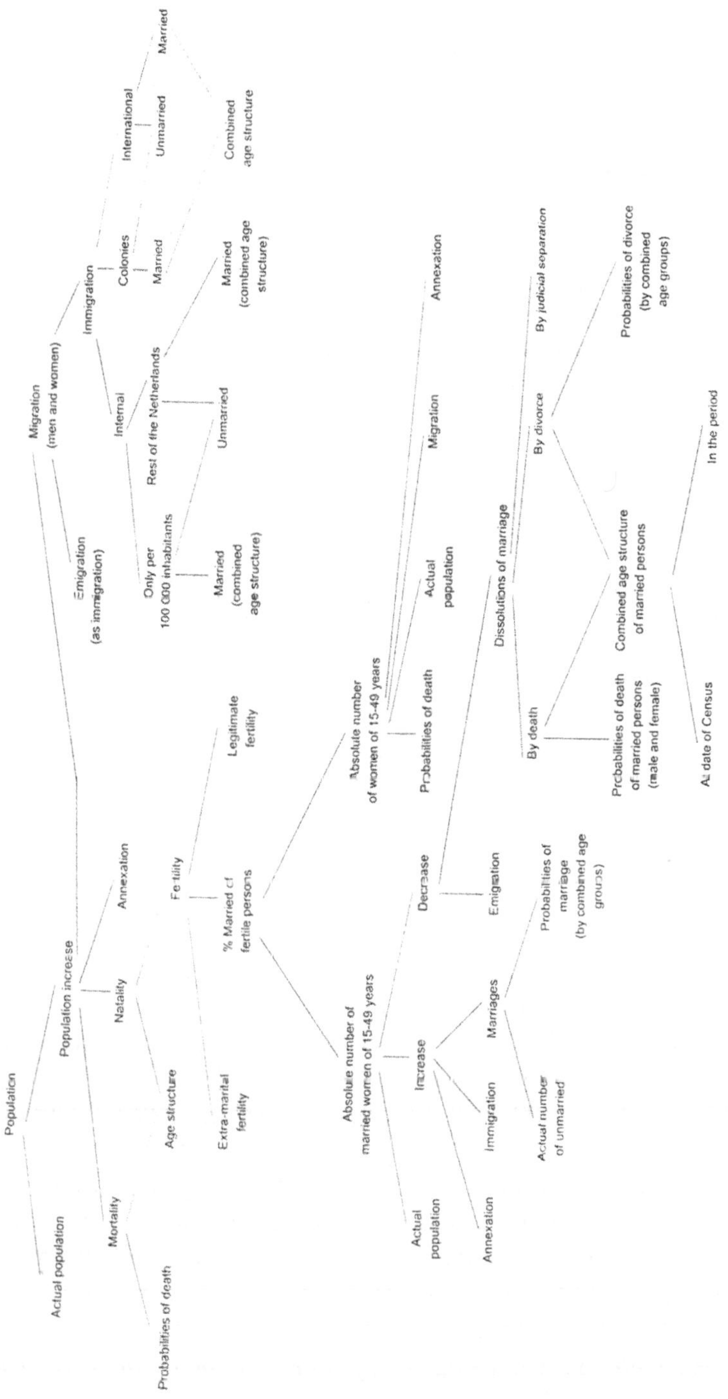

Figure 2.1: Diagram annexed to Wiebols' letter

is most in line with the specification mentioned in my thesis. [...]

The specification concerns the following factors (the parts that are underlined refer to men and women, the parts that are not underlined refer to women only).

1 and 1a. Present size of the population according to age structure and marital state, and the specified age structure of married people according to the age of the male, at the date of the population census.
2 and 2a Probabilities of death (separately for singles and married people).
3 Extra marital fertility by age group.
4 Marital fertility by age group.
5 Probability of marriage by age group (specified by age of the male).
6 Combined age structure of married people.
7 Probability of divorce by age group (specified by age of the male).
8 Probability of separation from bed and board.
9 Migration.
10 Annexation.
[....]

On the base of my present knowledge of the problem, I believe that with the proposed modifications an estimation for Amsterdam can be made that is based on the specification mentioned, without the risk of having to make too many by-speculations. However, the execution will be very labourious.

I hope that I have been of any use to you [...]. I would be happy to give further explanations, either in writing or verbally in Amsterdam.

Yours sincerely, G.A.H. Wiebols.

6.9 million inhabitants in 1920. He was an outspoken realist in his interpretation of these figures: '*It may be clear by now that the possible future population size is to a high degree dependent of the future economic development. If the development does not allow the Netherlands to have a population of 12.5 million inhabitants in 2000, well then, in that case the future population size will be smaller. In that case the migration factor will act as a correction factor, or (and) the course of the natality and mortality graphs will be different from those which we have calculated. We do not know, what the future population size will be in reality. We can only draw the conclusion, that the Netherlands, if not hit by particular catastrophes in any form, if no migration takes place and if the fertility and mortality probabilities in the future follow the assumed courses, will count a population of more than 12.5 million inhabitants*' (Wiebols, 1925, p. 109).

It is no exaggeration to portray Wiebols as the originator of modern Dutch population forecasting. Instead of calculating the future size of total population on the basis of extrapolation of linear or geometric population growth (the current standard method) he introduced age structure and age specific mortality (life table mortality rates) and fertility. In doing so he turned population forecasting into demographic forecasting, completely transforming the methodological landscape of the Netherlands.

Moreover, Wiebols' letters to Van Zanten form the apex of pre-war Dutch urban and regional population forecasting history. Almost at the very moment of the introduction of true demographic forecasting, they highlight the direction of future modelling of this new and innovative method in population forecasting, only to be achieved in the Netherlands in the second half of the 20th century. Had the letters been published, or his ideas fully implemented, his fame would have been greater. Nevertheless, his influence on population forecasting methodology in the Netherlands has been considerable and lasting.

2.3 | Wiebols' Motives

In the 1920s, progress in population forecasting methodology was prompted by interest in the population problem. In the Netherlands, progress resulted from growing anxiety concerning the consequences of rapid population growth in the years following the First World War. This concern had also been the incentive underlying Wiebols' thesis. Not only did Wiebols innovate Dutch population forecasting methodology; his forecast stimulated the national debate on the population problem in the inter-war period by providing it with a sound and convincing quantitative foundation. It is interesting to note from his letters to Van Zanten, that, in his view, forecasts of the populations of the big cities were the means of getting a better population forecast of the Netherlands as a whole. Amsterdam had more specific statistics available than the Netherlands Central Bureau of Statistics (NCBS). In Wiebols' eyes, statistical data with better specification could lead to more accurate population forecasts. That municipal forecasts could serve other purposes was, in 1925, beyond the scope of his interests.

2.4 | Wiebols' Failed Career

The duration of Wiebols' active contribution to the development of demographic forecasting was brief. He failed to make a professional career in this field. Although eager to find employment, he was surprisingly unsuccessful and had to be satisfied with jobs outside his main expertise.

On 13 March 1928 he applied in vain for a post that conformed with his standards and field of interest. A new Division of Town Planning and Town Development was created in the Department of Works and Public Buildings of Amsterdam. The main task of the Division was to prepare and design a new General Extension Plan. In town planning literature, great play is made of the guiding CIAM principle of functional separation of the General Extension Plan of Amsterdam. The planning principles underlying it are considered by modern physical planning theoreticians to be even more innovative. The first principle is that planning is more than design; the second is that surveys are necessary to provide a basis for the assessment of the requirements the plan has to meet (Faludi and Van der Valk, 1994). The analysis of future economic growth was believed to be of the utmost importance and the extension plan staff had to include an economist. As the director of the Department stated: '.. *I judge it necessary to connect a young economist to the division, in order to enable it to tackle all kinds of economic studies in the field, the results of which will be of importance to a well-considered judgement of the economic necessity or the consequences of measures or extension plans to be proposed'.*[12]

Wiebols was one of many applicants for the job. Definitely justified, he considered himself to be eminently eligible, as his letter of application indicates: '*...am I of the opinion that co-operators are needed, that have occupied themselves in the field of economics and statistics. To a high degree town planning will be influenced by expectations, both for Amsterdam as for other towns, regarding the future size and composition of the population, migration influences, economic developments, relocation of markets et cetera, expectations that by economic and statistical research only can obtain a somewhat solid foundation. Because of the writing of my thesis, which you will find enclosed, and because of the fact that as a co-operator I have joined a study club for town planning issues, that was established by Mr. Van Lohuizen*

[12] Director De Graaf to the Alderman of Works and Public Buildings, dd. October 27, 1928 (GAA, Arch. 5180/1928/1538).

and professor Granpré Molière, in combination with 3.5 years of practice in another field, I think to be able to do a good job in the division that is going to be established.' [13]

Presumably Wiebols was devastated not to be selected. Instead, G.T.J. Delfgaauw, an economist who had recently completed his studies at the Municipal University of Amsterdam, and another member of the study club for town planning issues, was appointed.[14] One can only guess why Delfgaauw, who was definitely qualified for the job, was preferred to Wiebols. Presumably there were several reasons. Perhaps Wiebols was considered too old; at the time he was 33 years of age, while Delfgaauw was ten years his junior.[15] That Wiebols was eager to obtain the position can be concluded from his readiness to be content with an excessively low salary of 2,600 Dutch guilders, about the same amount as that paid to Delfgaauw.[16] The head of the new division, Scheffer, received a yearly stipend of 10,000 guilders and Van Lohuizen, the town planner who would have been his immediate superior, received 7,500 guilders (Hellinga, 1985, p. 35).

Perhaps Wiebols was not selected because of the very fact of his national forecast. The idea was cherished that future population resulted from the future development of the labour market of Amsterdam. Consequently it was thought that priority had to be given to an economic forecast of Amsterdam. It was anticipated that the future labour-market could be deduced from such a forecast and that a calculation could be made of the number of households the labour market would support. Taking into account assumptions of the number of commuters working in Amsterdam, the future population of Amsterdam could then be deduced.

A mere demographic forecast was seen as a second best solution to the problem. Doubts may have arisen as to whether the economist-statistician Wiebols had developed into a demographic forecasting expert rather than the economic forecaster they sought. In contrast, in 1928 Delfgaauw wrote that

---

13 GAA, Afd. PW 1928, nr. 19662. Professor Granpré Moliere was an architect-town planner at Delft Technological University.

14 On the membership of Delfgaauw of this study club, see interview R. Bruins (NAi: Arch. Van Lohuizen, inv. nr. d.25).

15 List of applicants (GAA, Arch. Publieke Werken, doss. nr. 19662).

16 Application letter to the Director of Works and Public Buildings, 13 March 1928 (GAA, Arch Publieke Werken. 1928. doss. nr. 19662).

'.. *only an analysis of the structure of our big towns, of the function which they perform in the economic organism of the Netherlands can answer the question of the degree of the attraction which is exerted on the 'surplus' population'* (Delfgaauw, 1932, p. 9). This was one of the conclusions of his report of an investigation into the extent to which efforts to achieve a decentralisation in the settlement of the population could be based in the location tendencies of trade and industry. The aim of the study was to investigate the present and future part played by the large cities in national population growth.

Delfgaauw undertook his study while at the Netherlands Institute for Housing and Town Planning (NIVS).[17] It afforded him the opportunity of studying the relevant literature on future population growth, including Wiebols' dissertation. Through it, he also became acquainted with two influential town planners of that period; Hudig, the director of the Institute, and Van Lohuizen, his future chief at the Division of Town Planning in Amsterdam. Van Lohuizen had been his supervisor while he was working at the Institute (De Jonge van Ellemeet, 1932). Hudig and Van Lohuizen were well acquainted with each other. In his report, Delfgaauw paid his respects to Van Lohuizen, saying that all he had done was to build on Van Lohuizen's earlier research (Delfgaauw, 1932, p. 10). Presumably, Delfgaauw had established better relations than Wiebols with the right people at the right time.[18; 19]

Ironically it was given to Delfgaauw to be the first to apply Wiebols' forecasting method at the level of a municipality and in doing so to contribute to the modernisation of urban and regional population forecasting (see Chapter 7). He made a forecast of the future size of the population of Amsterdam based on the guiding principles of Wiebols' national forecast when an economic forecast of Amsterdam was considered to be too ambitious a project. His forecast, made under the supervision of Van Lohuizen, was published in 1932 (Grondslagen, 1932).For Wiebols, the application for the position in Amsterdam was a distressing experience. The result was that the greatest Dutch expert

---

17 The text of the report was finished in 1928 but it was published as late as 1932.

18 Delfgaauw's supervisor Van Lohuizen had investigated the economic structure and the future town planning development of Rotterdam; also the tendencies of concentration and de-concentration of the population of the Western part of the Netherlands in de period 1869-1920 (Van Lohuizen, 1925; Van der Valk 1990).

in modern population forecasting in the inter-war period, was lost to the profession.[19]

2.5 | Wiebols' Source of Inspiration

What would have happened if Wiebols had been appointed in Amsterdam instead of Delfgaauw is a matter of speculation. Would he have seized the opportunity to try to put into practice the sophisticated ideas set out in his letters to Van Zanten, possibly in co-operation with the director of the Amsterdam Statistical Office as the provider of the required statistical data?

As appears from his book and letters Wiebols, knew that such an effort would have been very laborious. He was aware of the methodological details of working so rigidly with the many sets of age specific occurrence-exposure rates, but he was apparently quite optimistic about overcoming the difficulties of the task: *'There is the problem of how to apply in a workable situation probabilities that are dependent of each other. This is a technical problem that has been solved in the world of insurance, e.g. for probabilities of death and disability, and that could be solved in the same way for the population problem'* (Wiebols, 1925, p. 120). Nevertheless, he might well have become entangled in the complexities of modelling and calculation.

Wiebols proved himself to be an eager and enthusiastic student of professor A.O. Holwerda, his supervisor and a visiting professor of statistics and

[19] From an interview with R. Bruins, a member of the staff of the Division of Town Planning and Town Development (NAI, arch. Van Lohuizen, inv. nr. d. 23). After an interlude on the newspaper *De Telegraaf*, where he was responsible for the stock list, he worked until his death in 1960 at the Ministry of Economic Affairs in The Hague. Wiebols was not only an innovative forecaster but also in economics. In 1933, Wiebols published *Waarde en Prijs* (Value and Price), in which book he criticised contemporary mainstream economic theories in the Netherlands. Economics was then dominated by the Austrian School of Economics, a school characterised by rigorous subjectivism. Nowadays, Wiebols' criticisms would be considered part of mainstream economics, but in his day he was an economics dissident. Given Wiebols' worth as an economist statistician in population studies, one would have expected his criticisms to have been taken seriously. His contribution never became the subject of serious debate among economists and it has since been completely forgotten. The disregard on the part of C.A.Verrijn Stuart, one of the most prominent adherents of the Austrian school, was particularly remarkable. Verrijn Stuart had praised Wiebols achievements in the field of the study of (future) population (De Gans, 1995a; Plasmeijer, 1995).

insurance science. As is shown in Chapter 3, he was one of the few early representatives of mathematical statistics in the Netherlands. He was an advocate of the use of age-specific sets of risks of exposure to demographic phenomena into forecasting methodology (see Chapter s 4 and 5; also De Gans, 1995a). Wiebols was the first to put life table survival probabilities into the practice of population forecasting. As far as fertility was concerned Wiebols had to content himself with the use of general fertility rates as long as age specific fertility rates or age-specific marital and extra-marital fertility rates were lacking in Dutch statistics. His letter to Van Zanten a few weeks after his graduation describes him embarking on a much more rigid and systematic endeavour of working with sets of occurrence/exposure rates.

2.6 | Wiebols' Blocked International Reputation

Van Zanten's conception of statistics was very different from Holwerda's. As director of the ABvS, Van Zanten's main responsibility was the organisation and publication of objective and reliable public statistics relevant to science, policy making and everyday forecasting practice. In Van Zanten's view, statisticians and statistical offices should refrain from all action that could undermine people's confidence in public statistics. He was afraid that forecasting, because of its inherent essence of speculation, would prove detrimental to the statistician's main task.

Van Zanten did not deny the importance of population forecasting. In his opinion, it was the task of the statistician and the statistical office to provide the forecaster with objective statistical data and advice on how to use them. His letter to Wiebols, in which he asked for further specifications in order to prepare for the statistical allowances for the innovation of forecasting methodology in Amsterdam, is proof of that. New developments were followed closely and with an open mind. The letter is a demonstration of his awareness of the specific responsibility of the Bureau of Statistics for good and up-to-date population forecasting in Amsterdam. By keeping track of the developments and requirements of the field he himself became a passive population forecasting expert. As such he always felt it necessary to emphasise the speculative nature of forecasting. Van Zanten's view was that of the true director of one of the main statistical offices in the Netherlands.

Vignette 1: Allard Othmar Holwerda (1887-1944)

Holwerda was Wiebols' supervisor at the Business school of Rotterdam for his PhD thesis on the future size of the population of the Netherlands. Born in The Hague in 1887, Holwerda spent most of his life in Rotterdam, where he attended the Erasmiaansch Gymnasium. Holwerda studied mathematics and physics at the University of Leyden where Lorentz and Kamerlingh Onnes were among his teachers. He obtained his doctorate with a thesis on *Frequency curves* in 1913, and joined the life insurance bank, the Eerste Rotterdamsche Maatschappij voor de verzekering van het leven, which merged a few years later with the Nationale Levensverzekering Bank, as an actuary. In 1917 he started to lecture on statistics and life insurance science at the Business school of Rotterdam. His inaugural oration was entitled *The scientific school in statistics*. A few years later he became a visiting professor. Holwerda was one of the few early representatives of the scientific school of mathematical statistics in the Netherlands in the first decades of this century. In his opinion, scientific statistics is the mathematical science of mass observation. Its domain is establishing what is regular and typical in mass observations, and changes in these observations. As an actuary for a life insurance company, he was familiar with and interested in (the calculation of) future probabilities of death.

In 1922 Holwerda made an impact in the 'Great Neo-Malthusian Debate' on the population problem in the Netherlands. He expressed the opinion that understanding present developments and predicting the future course of population required working with sets of age specific occurrence/exposure rates (to mortality, marriage, marital fertility). As long as such information was lacking there was no objective knowledge and all that was left was guesswork starting from subjective presumptions. Holwerda stood at the basis of modern demographic forecasting in the Netherlands. He set Wiebols on the track of working along the indicated lines (Source: De Gans, 1997).

Vignette 2: Jacob Herman van Zanten (1874-1944)

Van Zanten was first and foremost a statistician, as demonstrated by the motto of his *ex libris* (*'Statistics. Looking for proportions in quantities'*). He was neither a mathematician nor a statistician by training, having studied law. He became involved with statistics while working with Falkenburg, the first director of the Amsterdam Bureau of Statistics (ABvS; the first of its kind in the Netherlands, founded in 1894), on an overview of charity institutions in the Netherlands (completed in 1897). Van Zanten's interest in statistics was evoked whilst working on his dissertation on the methodological aspects of the description of poor relief. In 1917 Falkenburg was succeeded by Van Zanten, who stayed in office until his retirement in 1934. As Director of the ABvS and as a prolific author on demographic and other statistical topics, Van Zanten took an influential position in the international community of (urban) statisticians and among students of population in the Netherlands.

In 1923, Van Zanten was appointed lecturer in statistics at the newly established *Faculty of Commercial Sciences* (from 1935 onwards, the Faculty of Economical Sciences) in the Amsterdam Municipal University. He was mainly interested in teaching public statistic, which caused the Faculty some problems in the 1930s when students complained of not being sufficiently informed about advancements in mathematical statistics.

Van Zanten was opposed to population forecasting by statisticians and statistical offices. In his view, population forecasting would damage the credibility of statisticians involved in such speculative activity. Under his direction the ABvS never became involved in population forecasting. In contrast, the 1937 edition of his textbook on statistical methods was the first in the Netherlands to contain a synopsis of state-of-the-art population forecasting, including references to relevant national and international literature. Consequently, Van Zanten was the first to introduce university students of statistics and economics to population forecasting.

Van Zanten was an influential member of the International Statistical Institute (ISI) and one of the first members of the Dutch branch of the International Union for the Scientific Investigation of Population Problems. In the period 1936-1941, after his retirement, he became the first Director of the Foundation for Population Research in the Reclaimed Zuyder Zee Polders. This was the first social science research institution in the Netherlands. Here, Van Zanten investigated the composition of the colonist population of the first reclaimed Wieringermeer polder. He demonstrated that, apart from the official criteria for the selection of colonists, another criterion had been at work; religion. In 1941 Van Zanten was evicted from all his posts in German occupied Holland because he was a Jew. He died in camp Bergen Belsen in 1944 (Source: De Gans, 1994a; 1997).

Dr. J.H. van Zanten and his Ex Libris

The alert reaction to presumed criticisms of the quality of the statistical data of his Bureau of Statistics in Wiebols' thesis provides further proof.

The liberal policy of the Municipality of Amsterdam gave Van Zanten the opportunity to be one of the very small number of Dutch statisticians participating in international statistical conferences, including those of the International Statistical Institute (ISI). The 1931 Session of ISI held in Tokyo holds a vital position in the intellectual history of modern population forecasting. Van Zanten was present in Tokyo. He had a golden opportunity to further the international propagation of innovations in forecasting methodology in the Netherlands, the general reputation of population forecasting in the Netherlands and in particular the work of Wiebols. He failed to seize the opportunity, as did the other Dutch participants. Why they failed to do so is attributed to the occurrence of a major methodological change in the study of future population. The question is addressed in Chapter 4. In any case, the consequence was that Wiebols' work, although its innovative quality and its utility at various geographical levels found ample recognition in the Netherlands, was unjustifiably not recorded in the international history of modern population forecasting.

2.7 | Reflections

The story of Wiebols' life as a demographic forecaster reflects the history of modern population forecasting in the Netherlands in the first half of the 20th century and places the reader in the mid-1920s in which modern population forecasting emerged in the Netherlands in its full glory as a subsidiary of the national debate on the population problem.

It is interesting to note the impact of the population issue on Wiebols' thinking. He saw population forecasts of the big cities as a means of studying the national population issue; the problem of a population that was growing too fast. Apparently, the practical problems facing municipalities were beyond his scope of interest. His approach to a forecast for Amsterdam set out in his letter to Van Zanten was a systematic theoretical elaboration of the directions given by Holwerda; quite ahead of his time, but too far from the everyday needs of urban population forecasting. That the task of the application and elaboration of Wiebols' method at municipal level fell to Van Lohuizen and Delfgaauw was a loss to demographic forecasting from a scientific point of view, but presumably a blessing for the advancement of urban and regional forecasting for policy making purposes.

Wiebols was one of the international pioneers of demographic forecasting and definitely the originator of modern population forecasting in the Netherlands (evidence justifying these statements is given in the next chapters). His story also demonstrates that the apparently best qualified expert for the job was not the person to propagate the methodological innovation, bringing it from national level to that of the region and municipality. Whether he was also the founding father remains therefore to be seen. His position in demographic forecasting is comparable with that of a biological father. He failed to develop into the social father, because in failing to obtain a post in the demographic forecasting and planning profession he lost the opportunity to continue to contribute to the development of the field. In the event, Van Lohuizen and Delfgaauw became the true social fathers of Dutch demographic forecasting in making Wiebols' method suitable for application at urban and regional levels, as described in Chapter 7.

It is often assumed that modern science is characterised by a fast and easy exchange of information of new developments. Wiebols' biography makes it clear that this was not so in the case of his innovative contribution.

3. The Emergence of Demographic Forecasting in Europe

3.1 | The Origins

Wiebols' demographic forecast did not come out of the blue. There are a few early occurrences of innovative predictions of future population in other countries that can stand comparison with many forecasts of the interwar period, albeit without Wiebols' methodological sophistication. Students of population forecasting history only agree about the time and place of the beginning of demographic forecasting if the focus is on methodology. In that case, England in 1895 is close to the mark with Edwin Cannan's first national demographic forecast.

Forecasting as such was well accepted as a useful social activity by that time. In many countries, projections of total population contributed to everyday decision making. Estimates of future population growth were made for a variety of purposes. In 1880, for instance, a long-term forecast of the total population of England and Wales, based on decreasing growth rates, was related to an estimate of future coal consumption and the moment of its exhaustion (Price Williams, 1880). A few years later Cannan remarked that there was not "*... a builder nor a town council in the country that is not obliged to prophesy every month what the growth of population in a particular district is likely to be*" (Cannan, 1895, p. 505).

The examples given all originate from England. They defy the assertion of Spengler (1936) that population forecasting in the years prior to the first world war was common only in the United States and France. It was common in

France, because of the fear that the country might lose its influence on the continent. France had seen a fast decline after 1850-1860 of the rate of population growth in comparison with England and Germany. In the United States it was common for opposite reasons. There, extrapolations of the observed high growth rates brought people to expect the United States to become the world's most powerful nation.

Spengler was mistaken with respect not only to England, but also to France. In France, interest in population forecasting diminished from the second part of the 19th century on. The spectre of overpopulation disappeared because of improved economic development and demographic evolution. Among French statisticians, reduced interest in forecasting resulted from the awareness that fertility was continuing to decrease rapidly. The usual interpretations of Malthus' law of geometrical population growth, based on the assumption of an unvarying rate of total population growth, proved to be false. It was no longer possible to estimate precisely the time for the population to double. Efforts in France were directed instead at the perfection of mathematical instruments and the analysis of the course of the demographic evolution (Lachiver, 1987). The consequences of shaking belief in the existence of a law of population growth were extensive. Jacques Bertillon, for instance, concluded bluntly that it was better not to meddle at all in efforts to predict the future (Lachiver, 1987; also Jean Bourdon, in Sauvy, 1932, p. 340, Discussion).[20] The consequence of the French attitude towards population forecasting was that France played no part in the innovation of forecasting methodology until the mid-thirties, when Sauvy (1928; 1932) made the first modern forecasts of France.

As long as it could be assumed that population growth was governed by laws such as the mathematical law of geometric growth, the future course of population and its doubling time were considered to be more or less certain. When belief in such laws had been undermined, probability and speculation replaced certainty. From then on the future was open. New models of demographic reality were sought to supply theories of the future evolution of populations and new methods of forecasting. As will be discussed in section 3.3. forecasters had a variety of population theories from which to choose.

---

[20] Jacques Bertillon (1851-1922) was head of the municipal statistical office of Paris and a demographer of international repute.

Increasingly, long-term forecasting was founded on empirical population theory and therefore on the statistical analysis of observed trends.

3.2 | Dutch Statisticians and the Schools of Statistics

Statistics provide the roots of modern population forecasting. The significance of statistics as a separate branch of science increased with the growing demand of statistical data. But differences in the position of statistics as a science between countries were considerable At the end of the 19th century, the scientific picture of statistics and demography differed considerably from that of previous centuries: In the 17th century, demography and statistics corresponded to political arithmetic, *'a great intellectual adventure within the general perspective of the mathematization of the world'* (Dupâquier, 1985, p. 393). Empirical data were either lacking or were state secrets. Therefore, great efforts were put into conceptualisation. The history of demography of that period is of a biographic nature: Scientific progress was marked by the contributions of great personalities such as Huygens, De Witt, Petty, Leibnitz and Halley.

The epoch of scholars was succeeded by that of technicians working in a positivist ambience. It was the time of the great 19th century demographers: Statisticians such as Quételet in Belgium, Farr in England, Bertillon father and son in France, Körösi in Hungary, Kiaer in Norway, Böckh and Lexis in Germany and Bodio in Italy, scientists capable of looking beyond the scope of their figures (Dupâquier, 1985). In his *History of Statistics* Westergaard (1932) described the scientific equipment of statistics on the threshold of the 20th century as fully sufficient to cope with an extension of problems. The collection of observations was much easier than before, technical methods were in rapid development and theory was in constant evolution: " ... *A critical sense was awake which allowed no statistical field to be untouched till no doubt as to the results was left*" (Westergaard, 1932, p. 272).

Population forecasting methodology was not left undisturbed by these developments. Political arithmeticians developed important tools of demographic analysis that could be used in modern demographic forecasting. Statistician-demographers of the 19th century added many more. Well before 1895, the most important tools needed for demographic projections were ready, even though the necessary statistical data were not yet available (United Nations, 1973, p. 52; Smith and Keyfitz, 1977, p. 193; De Gans, 1993, p. 18).

The increased availability of statistical data resulted in ever extending time series of numbers and rates of demographic phenomena. Current population theories could be tested and attempts to formulate new theories encouraged.

At the end of the 19th century, the course of statistical science in the Netherlands and surrounding countries diverged, with consequences for the development of population forecasting in the next century. Up to the middle of that century, two different fields of knowledge, indicated with the name of 'statistics', were running side by side with few connections between them. These two fields only showed signs of coalescence in the second half of that century (Kendall, 1970).

The true ancestor of the first field, modern (mathematical) statistics, is *political arithmetic*, not 17th century statistics, as one might expect. The last mentioned form of statistics was entirely concerned with the description of political states. It was the forerunner of modern public statistics and the genre of geographical and political works of reference that often contained immense amounts of interesting information, but were 'statistical' only in the sense of the *Staatenkunde* in German countries (Kendall, 1970; Stamhuis, 1989).

The foundations of both political arithmetic and public statistics were laid in the Italian city states as early as the 14th and 15th centuries. But there a statistical approach was lacking. Counting was by complete enumeration and tended to be a record of situation rather than a basis for estimation or prediction in an expanding economy. The turmoil of war in Europe meant that political arithmetic did not start to develop until about 1660 (Kendall, 1970). In the mean time, statistics in the sense of the German *Staatenkunde* continued to flourish; the political description of states became more and more numerical as additional data were collected, but remained essentially the systematic collection of facts.

One of political arithmetic's first and great achievements was the construction of life tables. The development of life table theory began after John Graunt (1620-1674) had published the first table in his *Natural and Political Observations upon the Bills of Mortality .. of the City of London* in 1661. Progress was rapid and the basic ideas seem to have arisen independently of each other at many points in Western Europe (Kendall, 1970).

The 17th century saw the true beginning of statistics. Anyone reading this century's scientists, wrote Kendall (1970, p. 46) "*... must feel that this is the point at which statistics really began. The whole attack of the problem can be appreciated by a statistician or a demographer nowadays. These men, one feels, thought as we think today, with most imperfect material but with enviable penetration. They* ***reasoned*** *about their data*". Through William Petty (1623-1687), who gave it its name, political arithmetic developed rapidly in the next ten years. It found a firm footing in the Netherlands where its development, in particular in life table construction, benefited from the contributions of many outstanding people: The Mayor of Amsterdam, Johan Hudde (1628-1704) and the gifted Dutch statesman and Pensionary of State Johan de Witt (1625-1672) in the 17th century. In de next century Nicolaas Struyk (1687-1769) and Willem Kersseboom (1691-1771) (Dupâquier, 1985). The famous mathematician Christiaan Huygens (1629-1695) and his brother Lodewijk can be ranked among the pioneers of life table theory. Eight years after Graunt's life table, the brothers worked out how to use the table to calculate the life expectancy. However, the scientific world could not profit from their findings because their correspondence was not published until 1920 (Huygens, 1920).

In demographic forecasting, the contribution of the Dutch actuary Willem Kersseboom is of special interest. Kersseboom was one of the most prominent statisticians of his time and is ranked as the first professional actuary. In his writings of 1738-1742 he advocated the idea that estimates of population totals could be made with the help of life tables if the yearly numbers of births were known. As late as 1833, Rehual Lobatto (1797-1866), a mathematical advisor to a Dutch insurance company and a prominent Dutch statistician, considered it appropriate to use Kersseboom's 1742 life table for the calculation of tariffs of female annuitants (Stamhuis, 1989). Kersseboom was convinced that the estimation of total population should be based on statistics of births rather than on statistics of mortality, because the age structure of a population changes under the influence of the number of births (Kersseboom, 1738-1742; also Westergaard; 1932; Dupâquier, 1985).[21]

[21] For an overview of early estimators of total population see Greenwood (1942/1970: 90-96).

In England, political arithmetic and public statistics coalesced in mathematical statistics, eventually giving birth to the *English school of mathematical statistics* which developed between 1880 and 1920. The journal *Biometrika*, founded by Karl Pearson in 1901, became its medium of communication (E.S. Pearson, 1967/1970). Through G.A. Yule's *Introduction to the theory of statistics* (1911), in which mathematics played an essential part, statistics began to be considered as a separate branch of science (Stamhuis, 1989).

In the Netherlands, the course of the history of statistics developed in its own distinctive way. Despite the rich Dutch tradition originating with Christiaan Huygens' inestimable contribution to probability calculus, mathematical statistics was almost unknown here until the 1840s (Houwaart, 1991).[22] Lobatto was the only person to consider mathematics important in the development of statistics in 19th century Holland. In the Netherlands, statistics coincided with public statistics for quite a while, resembling German *Staatenkunde* rather than political arithmetic. Lobatto took a solitary position in building on both the Anglo-Dutch tradition of political arithmetic and the French tradition of the calculus of probabilities. His attempts to evaluate statistical data with probability calculus found no further following after his death in 1866. If his line of thought was kept alive, it was not among the Dutch statisticians of the early decades of the 20th century such as Verrijn Stuart, Methorst and Van Zanten, but among actuaries in the world of life insurance (Stamhuis, 1989).

For a long time the scientific status of statistics in the Netherlands was neither determined by nor dependent on its mathematical content (Stamhuis, 1989). Initially, the connections between statistics and economics were very close, both in the Netherlands and in England. Economists and statisticians sought to improve prosperity and built on quantitative statistical data. But the relationship evolved in different directions. In England, the London Statistical Society (later renamed the Royal Statistical Society) evolved into a truly scientific society with statistics as its main focus. Its Dutch counterpart, however, the

[22] The calculus of probabilities was developed in France by Descartes, Pascal and Fermat (Dupâquier, 1985, p. 130). Huygens worked out the theory of probability calculus for himself just for the pleasure of it, as he and his brother had done with life table theory, and inspired by French examples. In his turn Huygens influenced French mathematicians such as Bernoulli and Demoivre after Van Schooten's translation (and publication) of his book (*De Ratiociniis in Ludo Aleae*; 1657). It happened to be the first book that was published on the subject of probability calculus (Kendall, 1956/1970).

Statistical Society, gave equal priority to both economics and statistics in the initial stages. In the end economics, not mathematics, prevailed. First, the society changed its name to Society for Political Economy and Statistics (VSS). In 1951 the *statistics* part of its name was given up. (Stamhuis, 1989; Mooij, 1995). Consequently, the scientific status of statistics in the Netherlands differed considerably from that in England and elsewhere.

An overview by Van Zanten of the prevailing opinions of statistics and its scientific status clarifies the Dutch position within the international field. Van Zanten distinguished four categories, which he summarised as *'statistics is no science at all; or statistical science is either the description by numbers, or the mathematical science of mass observation, or the doctrine of the method of mass investigation'* (Van Zanten, 1927, p. 16).

To the *'statistics is no science at all'* category belonged statisticians who thought of statistics as a method of research, and definitely not a science on its own account. Van Zanten regarded Verrijn Stuart as the main Dutch representative of this category. In the words of Verrijn Stuart himself: *'Statistics is not a science but a form of methodical bookkeeping of phenomena of life that are liable to mass observation'* (Verrijn Stuart, 1910, p. 8). His textbook on statistics excludes the mathematical approach of statistics because it is an application of probability calculus which belongs to the domain of mathematics, not statistics. For the same reason Verrijn Stuart excluded forecasting from statistics. In his view, statistics has to do with observation and not speculation about future developments (Verrijn Stuart, 1928).

Statisticians of the second category, mainly from German speaking countries, saw statistics as the quantitative description of mass observation in the form of tables (to be compared with the former *Staatenkunde*).

Mathematical statistics, the third category, enjoyed an increasing number of adherents who conferred on mathematics an important role in statistics. The statisticians of this category were of the opinion that statistics owed its scientific character to a mathematical way of thinking and to the mathematical analysis of the results of statistical research. Van Zanten reckoned among its representatives statisticians such as Edgeworth, Bowley and Yule in England; Liesse and March in France; Bortkiewicsz and Winkler in Germany and Austria; Mortara and Gini in Italy and Holwerda in the Netherlands (Van

Zanten, 1927). Some of these were to play an important part in the development of population forecasting and in discussions on methodology in the interwar period.

Lastly, there was the category of statisticians, among whom Van Zanten reckoned himself, who saw statistics as the doctrine of the statistical method (the method of mass research) and the science of statistics as the ordering of concepts, facts and theories regarding the observation of mass phenomena according to one system (Van Zanten, 1927). Van Zanten shared Verrijn Stuart's conviction that statisticians had nothing to do with forecasting. It was the task of statistics to provide public statistics that could be trusted. This trust should never be jeopardised by speculations about the future (De Gans, 1994a; also Chapter 4).

It is remarkable that Methorst, Verrijn Stuart's influential successor as Director of NCBS is not mentioned. It is not too wild a guess that Van Zanten assigned Methorst to the category to which he himself belonged, notwithstanding the fact that Methorst's view of forecasting differed from that of Van Zanten. It is a plausible assumption that Methorst's point of view remained constant throughout his professional life as Director of NCBS, where he was responsible for the provision of good and accurate official statistics. Methorst makes his opinion evident in a statement made towards the end of his career: '*The task of statistics is to collect, group, assimilate and to make public the observable phenomena, according to its own method (...). It is the truth mirrored in statistics. For a foundation of measures to be taken in the field of social-economy every civilised state should have the disposal of unassailable statistical data and should therefore be in the possession of a statistical apparatus in which all observations come together and from where information radiates*' (Methorst, 1943, p. 2).

Van Zanten's overview makes it clear that the school of mathematical statistics was not well represented among the leading Dutch statisticians of the first three decades of the 20th century. The only Dutch representative mentioned was Holwerda. Holwerda only started to exert influence in the Netherlands in the period after the first world war. He was an actuary with a keen interest in probability calculus and in statistics. His 1917 inaugural lecture as a visiting professor was on the scientific school of thought in statistics. Scientific statistics is '*the pure mathematical science of mass observation, which task it is to establish the regular and the typical in mass observations as well as*

Vignette 3: Coenraad Alexander Verrijn Stuart (1865-1948)

Prof. C.A. Verrijn Stuart, painted by M.V.A. Röling (1934)

Verrijn Stuart is characterised as the first modern statistician of the Netherlands (De Vries, 1948). Primarily he was a liberal theoretical economist who dominated the economics scene in the Netherlands for a long period. N.G. Pierson (1839-1909), the great Dutch economist of the second part of the 19th century and the driving force behind the foundation of the Statistical Institute (1884; predecessor of the Amsterdam Bureau of Statistics); the Central Statistical Committee (1892) and the Netherlands Central Bureau of Statistics (1899) was his teacher.

Verrijn Stuart was the first Director of NCBS. Verrijn Stuart followed Pierson in considering that the population in the Netherlands was growing too fast in relation to its economy and steadfastly maintained this conviction between the two World Wars. The necessity of a reduction in the birth rate dominated his views on demographic matters during the inter-war period. His attitude towards Malthusian theory and neo-Malthusianism was fortrightly positive, even more than that of his teacher, who held a more balanced view.

Verrijn Stuart was on equal terms with the leading international statisticians he encountered at meetings of the International Statistical Institute, of which he was Secretary General (1907-1911). He represented the Dutch government at many ISI conferences. Verrijn Stuart was a professor at the Universities of Delft (1906-1908), Groningen (1908-1917) and Utrecht, teaching political and theoretical economy. From 1891 to 1930 he was a secretary and a treasurer of the Dutch Society for Political Science and Statistics (VVS). For an even longer period (1896-1949) he was on the editorial board of the influential journal De Economist. He was the author of one of the first modern statistical manuals in the Netherlands giving ample space to the application of statistical methods to demography and the analysis of the course of population.

Verrijn Stuart's most important contribution to demographic forecasting was not his calculations. As a secretary and treasurer of VSS and an editor of De Economist, he kept the population problem on the Dutch scientific agenda. He set the stage for the innovation of population forecasting methodology in the Netherlands by providing the context for discussion and debate on the population issue.

Verrijn Stuart was not a person who could easily change his mind. He had a reputation of being conceited. This together with other less attractive personality traits may have cost him the presidency of De Nederlandse Bank, but that loss may have been to the benefit of demography and demographic forecasting (Source: De Gans, 1997).

Vignette 4: Hendrikus Wilhelmus Methorst (1868-1955)

Methorst was an amiable man, a distinguished organiser of public statistics and demography practitioners and the author of an excellent history of public statistics in the Netherlands (CBS, 1902). He was influential as a demographer of the Netherlands of the inter-war period, in the international fields of statistics and in the study of population. He was one of the most well known Dutch professionals.

Methorst completed his law studies with a thesis on labor colonies for destitute people in 1895. On the establishment of NCBS in 1899 he became its deputy director, a position he owed to Pierson. He succeeded Verrijn Stuart as Director of NCBS, where he stayed from 1906 to 1939.

In 1911 Methorst was elected Secretary General of ISI. He occupied this post until 1947. When plans were made for ISI to have a Permanent Office, Methorst succeeded (in close cooperation with Verrijn Stuart) in having it established in The Hague, winning the competition against such countries as Belgium and Switzerland. In 1913, when the Permanent Office of ISI was established, Methorst became its first Director, combining the function with that of Director of NCBS. It was Methorst's ambition to make The Hague the permanent seat of ISI. He succeeded in maintaining the Institute during the vicissitudes of both the First and the Second World War, when international statistical contacts were reduced to a minimum.

Over the many years in which he conducted the affairs of the Permanent Office, Methorst came to identify himself and his personal interests with those of the Permanent Office in such a way that he appears to have lost the distinction. At the end of his long career, when he was 75 years old, he was characterised as a person who had never tried to imbibe the Institute with any new ideas. He had just kept it going as it was in the 19th century, a fact which in reality meant that he in his later years had contributed to the decline of the influence of the Institute. Methorst must have found it difficult to keep abreast of developments of statistics. He seems to have played safe, keeping to the

Prof. H. W. Methorst

mainstream. Presumably, as he grew older, he was increasingly swayed by the distinction of his peers, directors of statistical offices, rather than by newcomers and innovators. For too long the tendency had been to elect members in recognition of distinguished past achievements; there had been a failure to recruit young, active new leaders in statistical development.

Methorst's view of statistics as purely descriptive, must have impeded an open attitude to the direction in which the science of (mathematical) statistics was evolving. ISI was at the end of the War in a 'deplorable position'. Although this could not be attributed to Methorst solely, he nevertheless was blamed as the main person responsible, because of the position he had taken for such a long time.

As the Director of NCBS, Methorst became the renowned organiser of Dutch public statistics. Both nationally and internationally, Methorst was a passionate advocate of the centralisation of statistics. He was convinced that statistics should never be influenced, neither in method nor in results, by the changing aims of politics and politicians. Methorst was the organiser of the demographic field in the Netherlands. After the establishment of the International Union of the Scientific Investigation of Population Problems in 1928, in accordance with the resolutions of the Population Conference of Geneva in the preceding year Methorst became the first president of the Dutch branch of the Union and the predecessor of Netherlands Demographic Society, also founded in 1928. He was a member of the editorial board of Metron, an international Italy based scientific journal, until the mid 1930s. He published in French, English and German, mainly on the topic of differential fertility in the Netherlands and on advances in the field of the development and organization of the Dutch population registration system.

Methorst's congeniality, organisational ability and strategic insights enabled him to position his adherents in important places. He was eager to enlarge his influence, both nationally and internationally, by furthering his status. In 1927 he acquired the title of Director General of NCBS. Presumably to enhance his status as a scholar, he gave courses at the Royal Military Academy at Breda, where he acquired the personal title of Professor (Source: De Gans, 1997).

in the changes of the observations' (Holwerda, 1917, p. 10).[23] Van Zanten could not help drawing the conclusion that in Holwerda's vision *'science begins, where investigation ends'* (Van Zanten, 1927, p. 15). Verrijn Stuart, Van Zanten and presumably Methorst also, all directors of statistical offices and the leading statisticians of the Netherlands, belonged to schools of thought that largely denied mathematics, and probability calculus in particular, a place in statistics. Verrijn Stuart and Van Zanten shared the opinion that statistics and speculations about the future had nothing in common.

3.3 | From Law of Population to Population Theories

The assertion, made in section 3.1, that 1895 marks the true beginning of modern population forecasting, is only justified from a methodological point of view. Had the focus been on theory and conceptualisation, the location in time would have been different. Theorising and conceptualisation of (future) population had been going on much longer. By 1895 some of the most important questions had already been formulated and answered (Hecht, 1980; 1990). Since theorising about future population is as old as demography, its beginnings have to be located somewhere at the end of the 17th century (Hecht, 1980; 1990; Kendall 1960/1970). Some early theories denied any structural increase of population. Others were derived from biblical concepts of the future and the Christian tradition. There were also early visions presupposing a future increase of population (Lachiver, 1987).

In the 19th century Malthus' concept of the course of population gained dominance. Malthus' basic proposition, his 'law' that population doubles itself every twenty-five years, remained influential far into the inter-war period. The

[23] Holwerda was not alone in the Netherlands. There were other scientists interested in mathematical statistics and probability, for instance Van Haaften, a specialist in life table theory, an actuary who was the first to hold a chair in life insurance mathematics. Also the astronomer J.C. Kapteyn; the mathematician M.J. van Uven who held a chair at the Agricultural University of Wageningen, and J. Tinbergen, the founder of econometrics (Duparc and Grootendorst, 1978, p. xxvi; Stamhuis, 1989, p. 43). Interest in mathematical statistics started to grow in the Netherlands in the 1930s. Students of economics at the University of Amsterdam, for instance, were dissatisfied with Van Zanten's one-sided courses of public statistics and asked for courses in mathematical statistics instead (De Gans, 1994a). Mathematical statistics gained general acceptance in the Netherlands after World War II (Stamhuis, 1989).

same holds for his propositions that population size is limited by the means of subsistence and that, if the poor would not voluntarily abstain from sexual intercourse and having offspring, nature (that is to say, the means of subsistence) would impose its own restrictions. The presumption that population growth is related to the means of subsistence, resources or human production underlies several other theories, independently of whether such theories stem from mathematical/biological or economic notions, such as the logistic growth and optimum population theory (United Nations, 1973; Bolte and Kappe, 1967). The assumption accounts for the close relationship between economy and demography in that period.

The Belgian statistician Quételet (1796-1874) took a mathematical approach that was different from that of Malthus. In 1835, he asserted that the resistance (the sum of obstacles opposed to the unlimited growth of population) increases in proportion to the square of the velocity with which a population tends to increase. In the absence of change in the underlying conditions, a population that has arrived at a certain point of saturation grows increasingly slowly. At Quételet's request, his compatriot, the mathematician François-Pierre Verhulst (1804-1849) looked for a fitting mathematical formula and arrived at a symmetrical *'S'- curve* as the most suitable function to describe the past and future course of population. The function was termed the '*logistic*' (United Nations, 1973; Dupâquier, 1985; also Chapter 1, f.n. 7). Contrary to Malthus' law, it had the assuaging quality of predicting that there would be an end to population growth: After a period of imbalance between birth and death rates, resulting in accelerated population growth, there would be a new equilibrium leading finally to zero population growth. Verhulst's contribution rapidly fell into oblivion. However, it was rediscovered in the early 1920s, when it conquered the scientific world for a while (see Chapter 4).

The economic *optimum population theory* was developed at the end of the 19th century. It was closely related to Malthusian and logistic theory. Like Malthus' theory, the optimum population theory was based on an assumed relationship between population and resources. The definition of the concept of 'optimum population' in a static sense is commonly attributed to Cannan, who worked on the concept in the late 1880s and early 1890s, although he did not give it its name (Robbins, 1927; Wolfe, 1929; United Nations, 1973). Cannan affirmed that "*at any given time the population which can exist on a given extent of land, consistently with the attainment of the greatest productiveness of industry possible at the time, is definite*" (Cannan, in Robbins, 1927,

p. 115). Exact estimations of the degree of under population or overpopulation at a specific point of time, however, is not possible and the existence of overpopulation or under-population is not susceptible of exact demonstration (Cannan, op. cit.).

At the end of the century, the impact of Darwin's theory of evolution on the societal and conceptual framework of population forecasting started to grow. The theory became one of the great new waves in 19th century science, comparable with Newtonianism in the previous century. The influence of Darwin's theory in other sciences went much further than its theoretical foundations could account for. Because of its prestige as a fruit of natural science (biology), the theory turned into a metaphor known as *social Darwinism,* taking on the proportions of an ideology. Social Darwinism was founded on the belief that social development, the progress of society, was determined by natural selection through the struggle for life (Noordman, 1991/1992). The influence of the Darwinian mechanism was 'polymorphous', because it appealed to opposing political movements, running from left to right wing and from anti humanitarian to humanitarian. Social Darwinism was used as an argument against socialism and social legislation, against social welfare and medical care for the poor and the weak in order to legitimise pure individualism, as an excuse for aggressive nationalism, and as an argument for the dominance of the white race. It was used by socialists, too, as a social theoretical foundation of the historical process of class struggle, although partly for strategic reasons: Monopolisation of the new biology by the opposite camp had to be prevented (Van Praag, 1977; Noordman, 1989; 1991/1992).

There was also a link between social Darwinism and the *eugenic movement* Some social Darwinians, including the father of the eugenic movement, Darwin's cousin Frances Galton (1822-1911), believed that nature could no longer be trusted to take care of the improvement of the human race. They were convinced that this had to become the task of mankind itself. The course of demographic evolution led them to doubt the success of this operation and resulted in a quite pessimistic vision of society. The improvement of the probabilities of survival of people in poor health brought Galton to the conclusion that, instead of a process of survival of the fittest, mankind was degenerating. It worried him that fertility was no longer primarily a characteristic of the 'fittest' in society. At the turn of the century, the first to reduce the number of their offspring appeared to be the upper classes in society, not the lower classes as Malthus had hoped (Soloway, 1990; Noordman, 1991/-1992).

The uninterrupted decrease of fertility in most Western countries in the period between about 1870 and 1945 resulted in two related, but not identical, fears expressed by a heterogeneous group of writers, scientists and politicians. Within nations, an imminent eclipse of the elites was feared, because the lower classes appeared to be the most prolific. Between nations, there was a fear of miscegenation and 'pollution' of sub-fertile indigenous populations by immigrants from fertile nations (e.g. Methorst, 1914). Not everybody was enamoured by social Darwinism and eugenics. It met resistance from scientists critical of the limitations of their field of science and from religious orthodoxy (Teitelbaum and Winter, 1985; Noordman, 1989; 1991/1992; Soloway, 1990).

In the Netherlands, social Darwinism and the eugenic movement received relatively little support in comparison with countries such as England and France. The Netherlands had not been forced to endure humiliating experiences comparable with those of France under its crushing defeat in the Franco-Prussian War of 1870-1871, or of England in the Boer Wars. The French blamed the defeat of their country partly on demographic factors. France's long history of fertility decrease had resulted in small numbers of men in the military age classes. Apart from the creation of Bismarck's German Empire and the French Third Republic, another result of the war was that the population issue became a permanent part of the French rhetoric and, occasionally, of its politics (Teitelbaum and Winter, 1985, p. 18). The English were forced to cope with the traumatic aftermath of the inauspicious performances of the British army against a handful of South African settlers in the Boer War. This raised doubts about the physical condition of the British soldier and fostered the fear of the degeneration of the British race (Teitelbaum and Winter, 1985; Soloway, 1990). The Netherlands had not experienced similar challenges. At the turn of the century the Dutch population was growing faster than ever and in Dutch culture strong resistance to social Darwinism stemmed from both Roman Catholic and Protestant orthodoxy. Although there were some hard-boiled social Darwinists in this country too, warnings against extreme opinions prevailed (Noordman, 1991/1992).

3.4 | The Theory of Parallelism

At the turn of the previous century statisticians, economists, demographers and policy makers were mainly interested in the future course of **total** population. This is hardly surprising, not only because of the belief in the Malthusian theory of geometric population growth, which was generally accepted, but also because of a lack of long time series and detailed statistical data of the vital components of population growth. Gradually the situation improved. Population censuses were held at regular intervals in an increasing number of countries and population registration improved. Statistical offices and students of population were able to establish longer time series of the components of population growth. Theories about the future course of population became more empirical. Analyses of statistical data made it clear that the demographic context of national population forecasts for the various countries of Europe was different (*Figure 3.1*).

Figure 3.1. Birth rates and infant mortality rates, 1841-1904. Averages per year of each decade

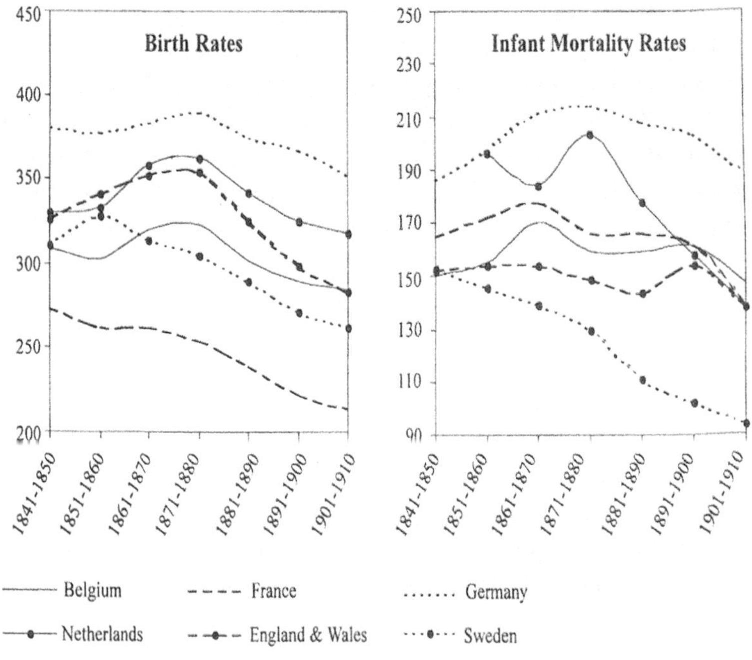

Source: Saltet and Falkenburg, 1907, p. 5, Tab. IV.

From about 1880 onwards there was a continuous decrease of birth and mortality rates in many European countries and in the United States of

America. A difference in the pace of mortality and fertility decrease resulted in an increase of natural growth in the period 1841-1900. Except for France and, to a lesser degree, England and Wales, mortality decrease was far ahead of fertility decrease. Differences in stage, tempo and level of departure in national demographic developments led to different reactions in the various countries. As two contemporary Dutch a students of infant mortality in the Netherlands, the physician Saltet and the statistician Falkenburg (1907, p. 3) concluded: *'Stagnation of the surplus of births over deaths means in the long run a stagnation of the population. In the competitive struggle of nations a stagnation of the population is a danger that can be the cause of a defeat. For that reason the courses of both births and deaths drew the attention in countries with a very low surplus of births over deaths.'*[24] Fears of overpopulation could be appeased, while new fears with respect to the future of population and its consequences were kindled.

The increase of systematically collected statistical data resulted in theories of future population based more on empirical analyses. These theories in their turn influenced assumptions made in population forecasting. Gradually a quantitative framework was created that allowed for a focus on the methodological aspects and technicalities of population forecasting, such as modelling, assumption making and calculation.

One of the new theories was the *theory of demographic parallelism* (Saltet and Falkenburg, 1907; Landry, 1949, p. 532). It was based on time series of observed birth and death rates. Elements of the theory can be found in the work of Quételet and Achille Guillard. Quételet argued that the number of births was determined by the number of deaths.[25] Guillard (1799-1876) was the grandfather of Jacques Bertillon, the originator and the main protagonist of the theory (Bertillon, 1903). The theory was based on the time-series of mortality and natality rates in European countries. It presupposes a relationship

---

24 R.H. Saltet (1853-1927) was a physician and a professor of hygienics. He belonged to a new generation of the 'hygienist movement'. They fought for medical state supervision and for the reinforcement of medical-hygienic facilities for the benefit of the population at large and they sought for a new science of health on a bacteriological foundation (Houwaart, 1991). Ph. Falkenburg was the first director of the Amsterdam Bureau of Statistics.

25 "*Le nombre des naissances est réglé par celui des décès.... Une grande mortalité marche généralement de front avec une grande fécondité*" (in: Saltet and Falkenburg, 1907, p. 4).

between these two vital components of population growth in such a manner that the course of natality is dependent of the course of mortality; a decline of mortality causes a decline of natality.[26] Bertillon's theory met no general approval. The most thorough of his critics in the Netherlands were Saltet and Falkenburg (1907), who asserted that Bertillon's theory was at variance with the facts. Although they agreed that the birth and death rates of the Netherlands and surrounding countries had decreased continuously since 1880, they rejected his conclusion that the observed decrease was uniform. In their view Bertillon was mistaken, for instead of following a parallel course, the graphs of many countries tended to diverge. The result was an increase of the natural growth rate, with the exception of France and England. In these countries the graphs tended to converge and the natural growth rate decreased. The Dutch critics also argued that Bertillon had drawn too far reaching conclusions from time series covering a relatively short period of observation.

It is interesting to note the arguments Saltet and Falkenburg used, which were partly based on the application of recently discovered instruments of demographic analysis. First, they contended, there is always a possibility that what is observed over a short period of time holds only temporarily. Second, if two lines run parallel, each line may influence the other, unless the parallelism depends on other factors.[27] Third, Bertillon and others believing in a built-in equilibrium of natality and mortality were criticised for neglecting the possibility that cases of a mortality surplus may have existed in countries escaping statistical observation. In these countries graphs of birth and death rates could diverge in a direction opposite to that observed in most other European countries. History demonstrates that such situations have occurred: Once great empires are later scarcely populated. Lastly it was asserted that Bertillon had not used death and birth rates that had been standardised for age-structure (the Körösy-rates). According to Saltet and Falkenburg, this failure had affected his comparison of the French demographic situation with that of other coun-

---

26 In the view of its adherents, the theory implied that an increase of fertility would result in an increase of mortality. In the words of Bertillon (quoted in Saltet and Falkenburg, 1907, p. 5): "*Les naissances sont moins-nombreuses où la vie est plus longue et réciproquement*" and "*la mort mesure la vie, la vie mesure la mort*".

27 Bertillon was also criticized for not having taken into account the possibility that the observed decrease of mortality and fertility might have been caused by different and unrelated factors (Landry, 1949).

tries in a negative way.[28] If he had followed the proper analytical procedures, concluded Saltet and Falkenburg, Bertillon would have arrived at a different evaluation of the French demographic situation. However, despite these severe criticisms, the theory continued to inspire students of population. The theory even played a central part, albeit implicitly, in one of the most vehement debates on forecasting methodology in the history of Dutch population forecasting. The debate is discussed in Chapter 5.

3.5 | The Prehistory of 'the' Demographic Transition in the Early Forecasts Revealed

The turn of the previous century saw the emergence of yet another theory of future population, that of *demographic transition*. The growing awareness that the observed decline of fertility in most of the Western world was taking on a structural form gave a new impetus to the investigation of the socio-economic causes of the decline. In 1893 Billings presented an overview of theories with respect to the socio-economic causes that motivated American couples to practice contraception. These included an increased desire for items that formerly had been luxuries but now were seen as necessities; a desire to preserve or secure social standing through expenditures not related to child-bearing; a desire to increase the quality of children by spending more money on each child; women's desire to be independent and women's growing evaluation of housekeeping as domestic slavery (Hodgson, 1983). In France at about the same time (in 1890), Arsène Dumont developed his 'social capillarity theory', which asserts that an individual living in a stratified society tends to reduce fertility in order to improve his social position.[29]

It took some time for the new knowledge to be generally accepted. In the Netherlands, the economist Pierson hesitated to take the consequences of the new knowledge seriously: *'Even if it were proven that an increase of prosperity and civilisation results in a decrease of the number of births, then it would*

---

28 Bertillon had asserted that death rates in France were more favorable than those of other countries. Saltet and Falkenburg could falsify the assertion by using Körösy rates. Joseph de Körösy (1844-1906), director of the Budapest statistical office, had presented his method of standardized rates at the 1891 ISI Session of Vienna (Van Zanten, 1927).

29 *"Le progrès de la natalité est en raison inverse de la capillarité sociale"* (Dumont, 1890, p. 130). More about Dumont's theory in Spengler, 1938/1968; De Jong, 1946; Hodgson, 1983.

not yet be proven that this cause would result in the decrease of population growth, for increase of prosperity and civilisation has a decreasing effect on mortality as well' (Pierson, 1890/1913, p. 175).

Pierson's reluctance did not alter the fact that, in other countries, students of population continued to investigate the empirical relationships between fertility levels and socio-economic factors. These investigations made them aware of the fact that fertility level is inversely related to three social economic factors: Standard of living; social class, and urban residence (Hodgson, 1983).

Until recently it was thought that, in 1929, Warren S. Thompson was the first to establish the relationship between fertility levels and socio-economic factors in a comprehensive way, although his work remained almost unnoticed by his contemporaries. Thompson specified three groups of countries with different rates of population growth. In 1934 Landry developed in his *La Révolution Démographique* the same basic ideas as Thompson though he did not appear to be familiar with his work. Landry postulated three stages of population development: Primitive, intermediate and contemporary, roughly equivalent to Thompson's three groups (Kirk, 1996).

Later, in 1945, Frank Notestein and Kingsley Davis transformed the relationship into *demographic transition* (Szreter, 1993; Kirk, 1996). The concept they called demographic transition was never used in a uniform way. Rather confusingly, it has sometimes been invoked as a theory (the theory of 'the' demographic transition); sometimes as a historical model, in attempts to formulate a generalised explanation of the process of mortality and fertility decline in Western countries; and sometimes as a predictive model, particularly after World War II, when it became popular as a theory that could anticipate the future trends of countries in the early stages of transition, or as a mere descriptive term (United Nations, 1973; Chesnais, 1996; Szreter, 1993).

However, it is less well known that, before demographic transition took on the form of any of these concepts, it had already been predicted in a very imaginative way by Harald Westergaard. The early innovators of population forecasting stand at the forefront of the intellectual history of modern demographic forecasting. They can also justifiably lay claim to priority with respect to the empirical foundations of the transition theory. In the following sections it is shown that the beginning of modern demographic forecasting coincided with the 'prehistory', if not the very beginning of the intellectual history of demographic transition.[30]

By predicting demographic transition, the first innovators of population forecasting methodology contributed to an early understanding of it. As many other contemporary students of population, whether influenced by Malthusian, logistic or optimum population theory, these forecasters presumed that in the long run there had to be an end to further population growth. The novelty of their respective contributions was that they integrated the observed decline of fertility into assumption making. At the same time they added to the plausibility of the assumption of a continuation in fertility decline. They made it seem credible that the end of population growth would be achieved in a non-violent way (Cannan), that the final situation was relatively open (Fahlbeck), that the period of fast population growth was transitory and that the future would see a demographic transition with far reaching consequences (Westergaard). Their endeavours, discussed in section 3.6, are astonishing examples of how new insights (as the insight into future demographic transition) can be arrived at just by using a forecasting methodology that fitted a new conception of the reality of population dynamics, that is to say by a re-specification of the population system.

3.6 | The Conceptualisation of Population Dynamics in the Futures of Cannan, Fahlbeck and Westergaard

By 1895 censuses were already providing relatively up-to-date information on the size, age and sex structure of the population of many European countries at regular ten year intervals. In that year, for the first time in history, census data of the population structure by age were used for forecasting purposes. The forecaster was the English economist Edwin Cannan (1861-1935), the forecast was of the population of England and Wales and the horizon of the forecast was 1991 (Cannan, 1895). In the intellectual history

---

30 The term 'pre-history' is used in Szreter (1993, p. 694 n. 16). In European tradition the history of demographic transition commonly originates with Alphonse Landry (1909). In recent years his name together with several other early precursors of demographic transition theory was added to the predominantly American intellectual history of the demographic transition (Hodgson, 1983; Szreter, 1993, p. 694 n.16).

of demographic forecasting methodology he stands at the head (Wolfe, 1928; Röpke, 1930; Hajnal, 1955; Smith and Keyfitz, 1977).[12]

Cannan's interest was far removed from statistics and even further from mathematical statistics. He was a theoretician of optimum population and like many contemporary economists interested in the problem of population. In contrast with most of his colleagues he refrained from contenting himself with mere calculation of the doubling time of population with geometric population growth methodology. He launched several fierce attacks on orthodox Malthusian and other forecasting approaches. It was a forecast of the population of London by the Royal Commission on the Water Supply of the Metropolis of London that induced him to question orthodox methodology (Cannan, 1931).[31] The combination of being a man of independent mind and means, a 'no nonsense' economist and a common sense pragmatist, helped him to follow untrodden paths (Samuelson, 1977). It is surprising that, with the exception of his paper of 1895, he kept away from mathematics and graphical presentations of data for the rest of his life. He had ventured into the field of population forecasting only one year before, in 1894. Here he demonstrated his sense of a pragmatic approach to population forecasting issues for the first time. He addressed the arbitrariness of administrative boundaries in municipal forecasting. Because administrative boundaries rarely coincide with the real urban areas, he advocated a regional approach to urban population forecasting (Cannan, 1894).

A non-violent end of population growth (Cannan, 1895)

The novelty of Cannan's contribution to forecasting methodology was threefold. Instead of mathematical (geometrical) extrapolation, he introduced a demographic forecasting approach, based on age, fertile age groups and survival proportions that –under certain conditions– are close to survival probabilities in life tables. His method was based on cohort-specific 'proportions of people surviving and living in England and Wales' calculated from successive population censuses.[32] He did not use age or cohort specific fertility

---

[31] Cannan did not stand alone in his criticism of Malthusian geometrical growth theory. At about the same time Pierson, in the Netherlands, interpreted Malthus' theory as being meant as merely a demonstration of how fast the multiplication of mankind can proceed if it is not hampered by anything (De Jong, 1946).

[32] Cannan used a rather confusing terminology: The 'survivors in England' were calculated with the "*rate of mortality and loss by migration*" (Cannan, 1895, p. 510).

Figure 3.2. Cannan's diagram on the population of England and Wales, 1851-1951.

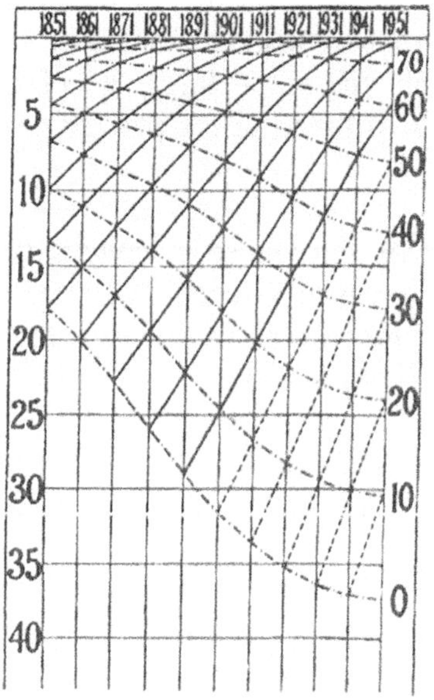

"The population in each census year is measured down from the top, the figures on the left indicating the number of millions. The lines sloping downwards from left to right divide the population at each census into persons over 80; persons between 70 and 80; between 60 and 70; and so on down to between 0 and 10. The ages are shown by the figures on the right. The lines sloping upwards from left to right divide the persons born in each decade from those born before and after it. They thus show graphically how each generation becomes smaller as it passes from under ten years of age at one census to over ten but under twenty at the next, and then to over twenty but under thirty, and so on."
Source: Cannan, 1895.

rates as such, but he was very close to it. The plausibility of his assumption of constant absolute numbers of births in future ten year forecasting intervals (up to 1990) was based on the observation of the tendency of decreasing general fertility and on calculated future changes in the number of women in the fertile age groups resulting from the numbers of live births two or three decades earlier.

From an analysis of the time-series of the number of births in the period 1853-1891, of the number of persons of age group 20-40 at successive census years and of the general fertility rate of that age group (men and women taken together) Cannan assumed that the decrease of general fertility in the period 1895-1921 would be such that the absolute number of births in successive future ten year intervals would remain constant at the level of 1881-1890. He assumed that the economic and social forces which had brought about the diminution of the number of births in the past 20 years would continue to operate in the future.

In the second place Cannan introduced a *'comparative state'* approach to forecasting based on the reconstruction of observed past and extrapolated future survival-in-England proportions in ten year cohort groups from successive population censuses (*Figure 3.2*).

Basically, the approach contained two innovative elements: A cohort approach and a diagram used as an illustration of the cohort dynamics in population forecasting.

By the force of his demographic model –and contrary to what was generally believed in his time– Cannan was able to predict that a cessation of population growth was at hand; not by wars, epidemics or starvation, as many post-Malthusians and post-Darwinians believed, but *in a non violent way,* by force of mere demographic evolution. As he himself confidently asserted: "*(...) The line shown is a much more probable one than that which might be laid down by the 'official' method, and which would shoot through the bottom of this diagram between 1921-1931 and encircle the globe before the diagram was widened very many yards*" (Cannan, 1895, p. 514).

While statistics of past population movements, population censuses, methods of demographic analysis and life table theory were readily available, it was Cannan's genius to redefine the population system that was to be forecast, to describe its past behaviour in a new way and make the proper selection of facts and tools for his forecast: "*The truth is that every estimate of population, past, present, and to come, ought to be founded on a consideration of the factors on which the growth or decline of population is dependent - births, deaths,- immigration, and emigration*" (Cannan, 1895, p. 509). Cannan opened the door to an explanatory demographic forecasting approach. The system to be forecast was no longer a population total, but a demographic population and the dynamics of the system was no longer governed by a law of total population growth, but by age-specific components of change working on the age structure.

Vignette 5: The history of the Net Reproduction Rate (NRR)

Samuelson (1977) suggests that Cannan stood at the cradle of yet another methodology with a real predictive power, the *Net Reproduction Rate* (NRR), because of his reasoned assumption of stationarity of the future numbers of births. Cannan had related declining general fertility to increasing numbers of persons in the fertile age group .of 20-40. Samuelson brackets Cannan and Kuczinski, to whom NRR is attributed, together. For a long period of time the rate was termed the '*Kuczynski rate*', after Robert R. Kuczynski (1876-1947). According to his son, Kuczynski was convinced that he was the originator of the concept: "*Von meinem Vater wissen wir, dass ihm die Idee zur Berechnung von Fruchtbarheitsraten der Bevölkerung (womit ein Grund der modernen Bevölkerungstheorie gelegt wurde) beim Anblick einer offenbar grösseren Anzahl reizender Frauen auf einer Schiffsreise kam. Natürlich haben sich die Wissenschaftler auf solche Entdeckungen am Schreibtisch vorbereitet. Aber die Idee kam eben nicht am Schreibtisch. Und ähnlich steht es um den Entschluss, dieser oder jenes Werk nun wirklich zu schreiben.*" (J. Kuczynski, 1972, p. 105). Kuczynski was the advocate of its utility in the late 1920s and 1930s and the first to apply it. By doing this he could indicate which populations were to expect an imminent decrease of population. However, his originator ship of NRR is questioned. Generally, Richard Böckh, a Director of the *Statistisches Amt* of Berlin and the mentor of Kuczynski for a while, is taken as the originator of this concept (in 1879) as for instance, in Wiebols (1925, p 40). The international introduction of the NRR took place in 1890. However co-originator status of the English statistician and demographer William Farr is also suggested (Lewes, 1984; Dupâquier, 1985). Apparently the invention of NRR was 'in the air', as so often is the case with inventions. The very notion of the concept cannot be found in Cannan's article. Cannan was both in time and in notion one or two steps away from its invention. In later years Wiebols (1925) used NRR in order to demonstrate the advantage of working with age specific fertility rates instead of with general fertility in population forecasting. As far as is known he was the first in the Netherlands to calculate to what degree the Dutch population was reproducing itself (Wiebols, 1925, p. 40). Methorst (1923) was the first to calculate age specific fertility rates of the Netherlands in the period 1910-1913.

In his own words: "*It is consequently much more true to say that the number of births depends in the first place upon the number of men and women between certain ages. For practical purposes the ages of twenty and forty are sufficiently near the mark, and they are much more convenient than the ages for each sex which would have to be taken if perfect accuracy were required.*" (Cannan, 1895, p. 512).

Cannan's forecast was based on sound demographic argument. He demonstrates a clear understanding of the basics of population dynamics. It is likely that the leading statisticians, heads of national or municipal statistical offices, shared the main body of knowledge of demography and the instruments of demographic analysis. Cannan, an outsider in this respect, was however the first to mobilise it for forecasting purposes, although he took care to try to avoid the impression that he was making a prediction: He had no desire "*to stake my reputation as a prophet on the growth of population following exactly the line shown in the diagram (...)*" (Cannan, 1895, pp. 513-514). He was only prepared to assert confidently that his line was a much more probable one than the line which might be laid down by the Registrar-General's 'official' method". It did not save him from being accused of being a false prophet in the 1920s (see Chapter 4, section 4.5).

Cannan's forecast is a convincing demonstration of the inter-relatedness of the numbers of births and deaths, of natality and mortality rates, and the age structure of a population. For that fact alone he deserves to be seen as a 'prehistoric' contributor to the demographic part of the concept of demographic transition.

'No end to the drop of fertility' (Fahlbeck, 1905)

In 1907, the American mathematician Alfred James Lotka (1880-1935) started work on stable population theory and its links with real populations, giving a strong impetus to the formalisation of population dynamics (Dupâquier, 1985, p. 417; Smith and Keyfitz, 1977, p. 75). At about the same time at its two-yearly Sessions in London (1905) and Copenhagen (1907) respectively, the International Statistical Institute helped to set the stage for the presentation of two more innovative predictions of future population, by Pontus Erland Fahlbeck (1850-1923), a professor of political science at Lund university in Sweden, and Harald Westergaard (1853-1936), who was a Danish statistician, an economist and a social reformer. Westergaard's started his academic career as an university lecturer in political science and the theory of statistics in 1883.

From 1886 to 1924 he was professor of political science at the university of Copenhagen (Van Maarsseveen, 1992).

Where Cannan reacted to the flaws of orthodox forecasting methodology. Fahlbeck's starting point was social Darwinian fears. The very title of his paper, *The Decadence and the Fall of Nations* (Fahlbeck, 1905) demonstrates his totally different background and context. In 1903 he added fuel to existing fears by the publication of his study of the natality of the Swedish nobility in the period 1885-1894. He demonstrated that natality and marital fertility of the Swedish nobility were far below the general figures for the total Swedish population (Fahlbeck, 1903). In 1905 he extrapolated observed series of birth and death rates of Great Britain and Ireland in order to demonstrate the consequences of the drop of fertility in many European countries for future population growth (see *Figure 3.3*). Like Cannan he took care to avoid the impression that he had entered the prediction business, underlining the fact that his extrapolation of the graphs of the birth and death rates was merely a guess, "*un pronotic*" (Fahlbeck, 1905, p. 381). Falckbeck's paper was a continuation of his 1903 study and discussed the consequences of the main conclusions –that the Swedish elite had stopped taking care of its reproduction– in a much wider and more social-Darwinian perspective, that of the decadence and decline of civilised nations.

The main thesis of Fahlbeck's paper was that all great empires and civilisations had come to an end because the leading classes no longer reproduced themselves and became extinct. In Fahlbeck's view this was becoming characteristic of western civilisation. Many years later, on the occasion of ISI's 50th anniversary in 1934, Fahlbeck's paper was praised by Friedrich Zahn, the chairman of ISI, because it '*... demonstrated, after an interesting study of the decadence of civilised nations of classical antiquity, the diminution of the births and the decrease of marital fertility*' (Zahn, 1934, p. 80).[33]

Fahlbeck's report was motivated by his conviction that the decadence and fall of civilized nations in the past had been caused by 'an internal illness' ("*une maladie interne*") rooted in 'abnormal' demographic conditions. By abnormal

---

33 It is not surprising that the German Nazi statistician Friedrich Zahn was interested in the subject of the decadence of nations, given the political state of affairs in Germany about 1934 (e.g. Höhn, 1986). It appealed also to French demographers as Alphonse Landry, because Fahlbeck linked political with demographic decadence (Landry, 1949, p. 44).

Vignette 6: The use of diagrams

Cannan's diagram did not escape the attention of colleagues, in and outside England, who were clearly impressed by his diagram (De Gans, 1994b). Bowley (1935) used it as a model of graphic representation when he started to teach at the London School of Economics (BLPES, Cannan 1020, 31-7-1897). Cannan was not the first one to use graphic illustrtions for forecasting purposes. Price Williams (1880) went ahead of him. The very first to use graphs of demographic phenomena was Christiaan Huygens (in 1669). It was a graph of mortality and expectation of life as a function of age. It became known as late as 1920 when his correspondence was published (Huygens, 1920, vol. 6, pp. 15-538; Dupâquier, 1985, pp. 209-211). Graphical representation in statistics started in the 18th century (Royston, 1956/1970; Dupâquier, 1985). The use of diagrams of demographic and other phenomena saw a considerable expansion in the course of the 19th century. An exhibition of charts and diagrams was organised at the Ninth International Statistical Congress of 1876 (Report, 1877). For a more extensive overview of the history the use of diagrams in demography see Caselli and Lombardo (1990). About 1874 other kinds of solutions to represent demographic phenomena graphically were developed, by German statisticians as Knapp (1842-1926), Becker (1823-1896) and Lexis (1837-1914) and by the Dutch mathematician Verweij (1843-1892). Particularly the closely related diagrammatic constructions of Becker, Lexis and Verweij have contributed to the resolution of the problem of age, period and cohort analysis of demographic phenomena on a discrete time basis. Their contributions helped for instance to enlighten and to illustrate the problem of the double ranking of vital events. The conditions for a correct calculation of age specific probabilities of death had been established much earlier and independent of each other by the Dutch life table expert Van Pesch (in 1866) and by the Germans Knapp and Zeuner (in 1868). These conditions were in the first place the ranking of the population at risk by year of birth instead of by year of age and a double ranking of deaths by year of birth and by year of age. At the 7th International Statistical Congess of 1869 the double ranking was recommended by another Dutch life table expert Von Baumhauer (Landry, 1949, p. 211-212; Dupâquier, 1985; Stamhuis, 1989). It took till the 1950s before Roland Pressat 'reinvented' the *Lexis diagram* and made it suitable for application in demographic teaching. As to the findings of Verwey's PhD thesis at the University of Utrecht (in 1874): They were presented at the international forum of statisticians (*The Statistical Society*), published in the *Journal of the Statistical Society* (1875) and referred to in Westergaard (1932). Notwithstanding, knowledge of his co-originator status with Lexis was lost till it was rediscovered by Vanderschrick in 1992 (Verweij, 1875; Vandeschrick, 1992).

conditions he meant the decline of the frequency of marriages and marital fertility.

From a forecasting point of view, Fahlbeck's ISI report is particularly interesting because of the analysis of the causes and consequences of trends in crude mortality and natality rates and because of the graphic representation of the observed and extrapolated course of these rates.

Fahlbeck's aim was to question the complacent belief that Western civilisation, and consequently the 'civilised nations', would continue for ever and to prove that such a conviction rested on three closely related assumptions that could be falsified with demographic arguments. The first underlying assumption was the belief in the unlimited progress of science and therefore of a continuance of the human dominance of nature. This assumption would only hold as long as there was sufficient human 'stock' that could feed science, that is, as long as the population of the 'civilised world' would at least take care of its reproduction.

The belief rested on another assumption, namely that there was no end to the progress of public morality, exemplified by respect for the life of an individual, nor to moral autonomy, finding expression in what Fahlbeck called the 'theory of perfectibility'. This amounted to a continuance of the readiness of women to give birth to sufficient numbers of children to guarantee future population growth. In spite of what Fahlbeck called 'modern altruism' which advocated a further increase of such female altruism, he saw no guarantee whatsoever that such a moral improvement of humanity would occur. Fahlbeck thought it unlikely that women would decide to produce the number of children necessary for populations to increase. The number of children was increasingly falling victim to a woman's freedom of choice.

This left one final assumption: The expectation of a continuance of the enormous population growth of the last century. However, this expectation, so demonstrated Fahlbeck, also lacked a solid foundation. One only had to look at the example of France to see that even this belief was far from certain. Fahlbeck saw no reason for the positive balance of births and deaths not to come to an end, or even turn in the opposite direction. Existing differences between European countries in absolute (natural) population growth only resulted from differences in the stage of demographic transition; in reality all

Figure 3.3. Fahlbeck's diagram. Observed and predicted birth and death rates of Great Britain and Ireland

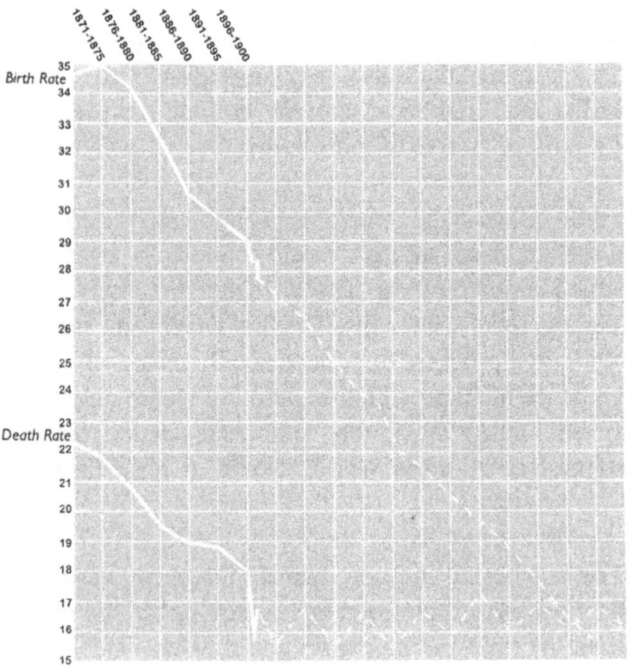

Source: Fahlbeck, 1905, p. 379

countries were on the same track on which France had preceded them. They were all about to witness a considerable decrease in natality and mortality. It was known from stable population theory that the level of the crude death rate in stationary populations is determined by the level of life expectancy at birth. The crude death rate of a population with a more or less constant yearly number of births, as was the case in Great Britain and Ireland, could only be lower than the rate determined by the level of life expectancy in a transitional period in which life expectancy at birth is increasing. Once the life expectancy figure reached its upper limit –and Fahlbeck expected this soon to be the case– the level of the crude mortality rate could only increase until it was in balance with the life expectancy again; from that moment on it would oscillate around a constant level.

Natality is, however, a different matter. Fahlbeck saw no natural lower limit to its level, apart from zero. Contrary to what was believed by Bertillon, the protagonist of the theory of parallelism, Fahlbeck was not convinced that in the end an equilibrium between births and deaths would prevail. There was no guarantee that demographic transition would result in a stationary situation in which the numbers of births and deaths would compensate each other.[34] Natality rates might quite well continue to drop after mortality rates had reached a constant level (Figure 3.3). The possibility of future negative population growth could not therefore be excluded: All depended on the will of mankind. The level of natality is determined by nuptiality and marital fertility. From his analysis of the prevailing social, economic and psychological tendencies, Fahlbeck could only observe signs of a continuance of the decrease of the rates of nuptiality and marital fertility. Population decrease instead of continuing population increase was imminent, for the crude death rate could not be expected to fall much further.

The lessons of vital statistics (Westergaard, 1908)

Fahlbeck lacked Cannan's pragmatism. His arguments were prompted by social Darwinian fears, including the fear of race suicide, but the demographic part of his argument was well founded. The assumption of a continuation of the decline of fertility and its consequences received further support in Westergaard's address, delivered two years later to the ISI Session of Copenhagen, bearing the title *The Horoscope of the Population in the Twentieth Century* (Westergaard, 1908).

Fahlbeck's demographic forecast based on the extrapolation of natality and mortality rates is interesting because he succeeded in escaping from belief in a final stationary situation. From the forecasting perspective Westergaard's contribution is even more interesting, even though the part related to the future lacks a quantitative backbone. Westergaard's demographic future is presented not as a real forecast, but as a speculation (as is clear from the use of the term 'horoscope') or, in present day terminology, a scenario. That he stressed the speculative aspect is not surprising when it is realised that his paper was presented before a forum of international top statisticians and directors of national and municipal statistical offices.

---

34 "*.. l'idée d'un pareil état d'équilibre est une pure hypothèse, dont la légitimité est fort contestable*" (Fahlbeck, 1905, p. 382).

Westergaard's scenario is less dominated by social Darwinistic fears. The future consequences of current demographic trends are treated more at a distance, similarly to Cannan's forecast of the population of England and Wales, but with no less skill. The prophetic quality of Westergaard's demographic scenario is astonishing and still striking by its 'modernity'. Westergaard gave an accurate prediction of the demographic transition of the 20th century and its consequences on a European and even a mondial scale, partly based on an intelligent and creative analysis of vital statistical data of Denmark. His horoscope is definitely based on a good knowledge of the dynamics of population change, the interaction between population structure and the components of population growth in particular. He had mastered the theory of life table populations thoroughly. He had a thorough knowledge of the demographic realities of his time, which were the point of departure of his presentation. His aim was to learn ".. *the lessons of vital statistics with regard to the future conditions of society.*" (Westergaard, 1908, p. 102).

Westergaard began with an analysis of mortality and fertility trends. From an analysis of Danish statistical data on the fertility of marriages with a duration of 10-15 years, he concluded that the practice of birth limitation had already taken up momentum and would continue at a rapid pace.

Two features of vital statistics were forcing themselves on the attention of Western nations: The increase of human life (that is, of life expectancy) and the decrease in the birth rate. According to Westergaard, statisticians agreed that the continuous decrease of mortality rates was structural and could not be interpreted as a momentary perturbation finding expression in a typical life table. The idea of a constant rate of mortality that had been lingering ".. *not only among Quételets pupils, but also in other circles of experts*" had proved to be wrong; no 'law of mortality' existed (Westergaard, 1908, p. 104). The immediate result of mortality decrease had been a growth of population in all Western nations such as had never been dreamed of in former centuries.

Mortality decrease had led to a population explosion in the Anglo-Saxon world in the second half of the 19th century, and consequently to a considerable change in the 'balance-sheet of nations', because of differences in their pace. The fall in mortality was followed by the decrease in fertility which began at the end of that century. Reduced birth rates could be observed everywhere, in Australasia, in America, in England as well as in Scandinavia and the capital cities of Europe. Here again the only marked difference had been in temporality: Some countries had retained higher birth rates somewhat longer than other countries. France was the only exception. Here a remarkable decrease had started at about the end of the 18th century causing almost stationary population growth because the increase of life expectancy had not been able to keep up with the decrease in fertility.

At the end of the 19th century net fertility was highest in the upper classes of society in spite of a lower birth rate than in the working classes, because the upper classes had a lower infant mortality. But that was changing rapidly. At the beginning of the 20th century, differences in infant mortality, for example, could no longer compensate for differences in fertility between the social classes. From his own analysis of Danish statistics Westergaard learned that fertility decrease was now taking place in all classes of society. He assumed that a similar development would take place in all European countries.

Westergaard predicted future shifts in the balance of nations from the differences observed in the pace of transition of mortality and fertility: "*Just as the Anglo Saxon race increased in numbers during the nineteenth century, so we may in future observe a quick increase of the Russians and Polonians until also this movement comes to an end*" (Westergaard, 1908, p. 110). These shifts would cause significant new changes in the balance of nations and America in particular would feel the consequences. In the past, Americans had been able to assimilate an influx of millions of mainly English speaking immigrants who had built a new English speaking nation. Assimilation in the future would be more difficult if immigration were to alter its character, with the mass of newcomers coming, for example, from Russia or Italy.

Westergaard cautioned against the views of those who consoled themselves with the expectation that the end of the transition process would see a return to old times: ".. *We shall not have the age distributions of former days, population will have an entirely different appearance, with its big numbers of old people and its relative small numbers of young persons*" (1908, p. 114). The change in age structure would have enormous effects. The burden of bringing up a child would be lessened, because more adults per child would be available to carry this burden. Everywhere in the active population on the other hand, "... *in offices and shops, the number of apprentices and juvenile clerks and assistants will be on decrease, whereas grey haired officials will be more abundant. And if it is true, that all new ideas are born in young brains, then this difference of age distribution is identic with a serious loss for the future population*" (Westergaard, 1908, p. 113).

His scenario predicted radical changes not only in the distribution of the population on a mondial scale, but also nation wide. Whilst in the 19th century the population had accumulated in urban centres, a spreading of the population in the rural districts could now be observed, resulting in the formation of many new settlements transforming the appearance of Denmark: "*Built with a curious minimum of taste as they are, after a uniform pattern, these new settlements have at all events great attraction on the surrounding population. Its is well worth noticing that this attraction at present seems much greater than that of the old centres of population*" (Westergaard, 1908, p. 111-112). In the future competition for immigrants between old towns and new settlements was to be expected, causing a reaction against further growth of cities and towns. The progress of communication (railways; cars) could lead to a decentralisation of population, as could already be observed in Copenhagen, where population was spreading from the inner areas to the suburbs and surroundings of the capital (Westergaard, 1908, p. 112).

3.7 | The First Indications of a Paradigm Shift

While in France a demographer of such international renown as Bertillon decided to abandon forecasting after renouncing belief in the existence of a law of population, Cannan, Fahlbeck and Westergaard dared to face speculation, searching for new ways of plotting the course of future population. They exchanged prediction within sound demographic frameworks for the certainty of a law. The predictions of these early innovators were based on thorough demographic analyses of past trends of the courses of the components of population growth. Current population theory was only accepted as long as it did not conflict with empirical demographic arguments.

Cannan's new approach was the most outspoken answer to his dissatisfaction with orthodox forecasting. In terms of forecasting methodology, Cannan's contribution was far-reaching; the plausibility of a cessation of population growth rested on quantitative demographic-analytic argument. Ten years later, Fahlbeck and Westergaard could build on new findings of demographic research and on longer time series of mortality and fertility. By then, the tendency of fertility decrease had become more prominent. Compared with

Cannan, Fahlbeck's forecast was less elaborate. It built only on a graphical extrapolation of the components of natural population growth. Presumably the most interesting part of his forecast was the openness of his '*demographic future*': He explicitly takes into account the possibility of a future deficit of the components of natural growth. As such it is a criticism of the theory of parallelism and of those later adherents to the theory of demographic transition, who presumed a more or less stationary situation in the final stage of the transition.

Westergaard's 'horoscope' gave the most comprehensive picture of the demographic future of Europe in the 20th century. It was a demographic scenario of future trends and consequences: The impact of differences in the pace of demographic evolution between countries on the balance sheet of nations, on the composition of the flows of intercontinental migration, on national urban-rural population distribution and its effects on the age structure and its consequences.

There are no indications that Fahlbeck and Westergaard were familiar with Cannan's study. Westergaard may have built on Fahlbeck's work, but there are no explicit references in his paper.[35] As endeavours in demographic forecasting the three contributions present themselves as unrelated events. In terms of demographic forecasting methodology there is no such thing as a straight line of progress: Cannan's quantitative approach (1895) was much more advanced than that of Fahlbeck (1905), while Westergaard's and Fahlbeck's forecasts were more advanced in terms of the analysis of fertility decline and the foundation of the assumption of its continuation. But each endeavour was innovative in its own way and deserves its proper place in the intellectual history of modern demographic forecasting.

It is by no means self-evident that these new explanatory demographic forecasts would be readily accepted. There are indications that prevailing beliefs and orthodox forecasting methodology inhibited their reception. Bertillon's reaction to the loss of his belief in population law is proof of that. Forty years later Cannan remembered that his forecast had encountered considerable criticism and he had not succeeded in convincing his contemporaries of 1895 that a

35 Fahlbeck and Westergaard were both members of ISI, elected in the same year, 1902 (The International Statistical Institute..., 1985). It is difficult to believe that Westergaard was unfamiliar with Fahlbeck's paper, as it was published in the *Bulletin* of ISI.

cessation of population growth was at hand (Cannan, 1931). However, his recollections might have been blurred after so long a period. Soloway (1990, p. 232), for instance, reveals that Cannan's forecast was cited in many projections made at the turn of the century when the fall in the birth rate became an issue of national concern. Fear of an absolute decline in population even caused a 'race suicide panic' in Great Britain at that time.

That it took about two or three decades before the new methodology started to gain ground is another indication of a reluctance to accept it. At first sight the hesitation could be explained from Cannan's outsider position: He was a student of economy, not a statistician-economist. He did not belong to the world of statisticians, even though he was a fearsome critic of the forecasting practices of the Registrar General's Office. The guardians of public statistics in England might have felt so harassed by his criticisms, that they refrained from referring to his forecast in ISI circles.[36] But that argument does not hold. Bowley, a member of ISI from 1902, acknowledged that as Cannan's student he had been impressed by his performance. He used Cannan's diagram in his lectures. It is unlikely that Bowley had any reason not to propagate Cannan's forecasting method in ISI circles if he saw it feasible to do so. Taking all the arguments together it suffices to conclude that Cannan met opposition to his forecasting method and encountered difficulties in combatting fixed opinions in scientific and professional circle – presumably because of the speculative character of his forecast, not his conceptualisation of population change.

Cannan was not alone in meeting opposition. Westergaard had similar experiences with statisticians and other forums of experts having difficulty in acknowledging the considerable decrease of mortality which had taken place in the 19th century: "... *The idea of a constant rate of mortality had been lingering for many years (...). It was not uncommon to expect an increase of mortality from other causes, when medical or hygienic progress had reduced the rates of a certain disease*". He recollected that the President of the *Institute of Actuaries* in London had chosen not to believe such a thing for ".. *if that be true any thing like a law of mortality would be at end*" (Westergaard, 1908, p. 104). Verrijn Stuart (1910) thought it useful to state explicitly that there is no such thing as a law of mortality.

---

[36] See De Gans (1994b).

If there was ever any belief in laws that govern the course of population, then it was faltering by the turn of the century. The efforts of the early innovators of population forecasting may have contributed to its decline.[37] Whether it was completely dead, or could readily be revived, is discussed in the following chapter. In the beginning period, utterances on the future course of population began to be considered as highly speculative. Cannan, for instance, made it quite clear that he had no desire to put his reputation at stake by prophesying that the growth of population would follow exactly the line shown in his diagram. He intended to show the directions of the development of future population; his confidence rested in the conviction that his line was more probable than the one laid down by the 'official' method (Cannan, 1895, pp. 513 and 514). Fahlbeck considered the graph of the future course of the crude birth rate as a mere conjecture, a prognostic.[38] Westergaard even classified his prediction under the category of horoscopes.

The first examples of demographic forecasting resulted from a new concept of the dynamics of population. From then on the future of population was seen as part of the broader context of social economic and demographic developments. This new concept of demographic reality clashed with the mainstream orthodox belief that the future development of total population was governed by a law of population. These clashes are tokens of the emergence of new times in population forecasting. It is tempting to say that they were the heralds of a change of paradigm in the study of future population: Fundamental beliefs, concepts and theories of population had to make room for others to allow for the emergence of methodological innovation.

[37] Whether such a rigid, dominant and generally accepted belief ever existed is open to question. Some sources contain information that seems to deny such an assumption (e.g. Spengler, 1936, p. 912 a.f.; United Nations, 1973, p. 559; Lachiver, 1987, p. 39). However, it is evident that a great number of authors worked from the presupposition.

[38] "*Nous posons ici la question de savoir quel aspect presentera dans l'avenir la courbe de la natalité. Nous l'avons reconstituée dans la diagramme ci-dessus d'après les expériences des dernières périodes. Mais il est clair que ce n'est là qu'une conjecture, un prognostic.*" (Fahlbeck, 1905, p. 381).

3.8 | The Reception in the Netherlands of the Early Endeavours in Demographic Forecasting

There are no indications of comparable Dutch contributions to the innovation of population forecasting prior to World War I. Were information and knowledge of the innovations by Cannan, Fahlbeck and Westergaard propagated in any way? The Dupâquiers (1985) hold that there was an easy exchange of new ideas between demographer-statisticians at the turn of the century. In their opinion, innovations became known within months, while discoveries took place almost simultaneously. If they are correct, one might expect to see a fast propagation of the new forecasting methodology; but that is not the case.

It took twenty to thirty years for modern demographic forecasting to take off. In terms of actual forecasting, the work of Cannan, Fahlbeck and Westergaard appears to have passed the Netherlands by. Until the end of the First World War only traditional geometrical and comparable extrapolation methods were used for the calculation of future population totals, although by the end of the second decade these calculations were founded on solid analysis of past trends and presumed continuations of the trends of the components of population growth. In Chapters 5 and 6 examples of this approach are discussed.

Outstanding Dutch economists such as Pierson and Verrijn Stuart even set the scene for an adherence to the old way of forecasting. In their publications they only presented calculations of the doubling time of populations with traditional geometrical growth methodology. This was not a result of a lack of expertise in life table construction; the Netherlands had a solid and excellent tradition in this field. Neither was it through a lack of expertise in demographic analysis, nor of a lack of knowledge of modern instruments of demographic analysis, nor of a lack of international contacts. Verrijn Stuart could stand international comparison in this respect and due to his position in ISI he was well informed about developments in the study of population and statistics (his manual of 1910 bears witness of this). It is more likely that his view of statistics stood in the way. Both demographic forecasting and mathematical statistics, formal demography included, were excluded from his field of interest. His views may account for the lack of innovation in Dutch population forecasting in this period.[39]

---

[39] The views expressed by the establishment of statistics may account, as well, for the fact that Verweij's contribution to the development of the 'Lexis diagram' left no trace in the

It remains to be seen how the predictions of Cannan, Fahlbeck and Westergaard were received in the Netherlands. Did they leave any traces? In line with Dupâquier's argument, one might expect a propagation of information of the new knowledge at the very least. People living at the end of the 20th century, familiar with the course of demographic developments in the past hundred years, are easily touched by the demographic insight of pioneers as Cannan, Fahlbeck and Westergaard and by the predictive power of their contributions. But what about their Dutch contemporaries? Were they able to grasp the prophetic and methodological quality? Were they ready to contribute consequently to its further propagation?

With respect to Cannan's forecast no traces of references in contemporary Dutch literature could be found.[40] This is surprising, for one might expect that the plausibility of his arguments, that the end of population increase would be arrived at in a natural and a non-violent way, would have been noticed by such fervent Dutch advocates of the necessity of a stop to future population growth as Pierson and Verrijn Stuart.[41]

It is a different case with respect to Westergaard's paper. Clearly, it was not meant to bear any official relevance to the Dutch nation. When Verrijn Stuart and Methorst, who had represented the Dutch Government at the ISI Session at Copenhagen, made their report of the proceedings of the conference, they omitted all reference to Westergaard's address (Verrijn Stuart and Methorst, 1907). However, this does not mean that the Dutch statisticians were not impressed by Westergaard. The paper was, for instance, referred to by Methorst as late as 1948 (Methorst and Sirks, 1948). That Verrijn Stuart was impressed by Westergaard's paper is apparent from a two page abstract of Westergaard's 'extremely suggestive' address included in his 1910 textbook on statistics. Having paid this tribute, Westergaard's scenario was dismissed

---

Netherlands (Vandeschrick, 1992; 1994). Even in Yntema's relatively recent *Inleiding tot de demometrie* ('Introduction to demometry'; 1977), which aimed at organizing the rather disrupted amount of literature on mathematical population theory within a systematic framework, references to Verweij are absent.

40 The first reference in a Dutch journal was by a non-Dutch author, Röpke (1930). The first –and more elaborate– discussion of Cannan's contribution by a Dutch author that could be traced, was in De Jong (1946, pp. 143-145). Cannan was better known from his contribution to optimum population theory.

41 It is difficult to believe that Cannan was unknown among Dutch economists. In 1904 for instance he was introduced to Pierson at the occasion of Pierson honorary doctorate at Cambridge university (Van Maarseveen, 1993, p. 345).

because of its *'purely speculative nature'* (Verrijn Stuart, 1910, pp. 286-287). Only much later, when longer time series of birth and death rates were available, was Verrijn Stuart inclined to mitigate his initial remarks by speaking of the *'somewhat speculative nature'* of Westergaard's paper (Verrijn Stuart, 1928, p. 344) and he considered the imminent change of the population by marital state and age and its consequences the most important part of Westergaard's prediction. How far Westergaard's lessons of vital statistics influenced Verrijn Stuart's rather one-sided opinions about future population is the concern of Chapter 5.

There is no evidence whatsoever that Westergaard's report had any influence on the development of forecasting methodology, or on assumption making in the beginning period. It is possible however that Wiebols was familiar with his report through Verrijn Stuart's manual of statistics. Moreover, in the absence of a solid quantitative framework, his contribution could easily be dismissed as being too speculative in nature.[42]

It is a different matter with respect to Fahlbeck's paper of 1905; it may have had a direct influence on the innovation of forecasting methodology in the Netherlands. The paper shared the fate of Westergaard's report of not being mentioned in the written reports of the conference proceedings of the Dutch representatives at the ISI Session of London (Verrijn Stuart, 1905a and 1905b) while on the other hand of being referred to in Verrijn Stuart's textbook. It is obvious that Verrijn Stuart took good note of Fahlbeck's forecast of mortality and natality and its consequences for the future course of the natural growth rate (Verrijn Stuart, 1910, pp. 285-286). Whether it incited him to improve his own approach of forecasting may be doubted (see Chapter 5). More than fifteen years later Methorst, Verrijn Stuart's successor in many ways, made quite an impression with a forecast of the total population of the Netherlands broadly similar to Fahlbeck's approach – based on graphs of observed and extrapolated birth and death rates (*Figure 3.4*).

It is highly probable that Methorst was familiar with Fahlbeck's paper. In several writings, most specifically in 1914 but also in 1919, Methorst gave

---

[42] See for instance Zahn (1934, p. 79). On the occasion of the 50th anniversary of ISI Friedrich Zahn, who had been president of the ISI since 1931, blamed Westergaard for *'.. having drawn no numerical conclusions for the future but only presumptions on future evolution'*.

a short treatise of the future course of the vital rates (birthrates and death rates) that appears to the reader as a concise and accurate summary of the essence of Fahlbeck's forecast, including its open future without explicit references to Fahlbeck (Methorst, 1914; 1919). Later, in his own forecast of 1922, which was based on extrapolations of the observed time series of birth and death rates, Methorst's demographic future became less open than that of Fahlbeck and more influenced by demographic parallelist notions (Methorst, 1922). On the other hand Methorst went one step further than Fahlbeck in not limiting himself to the extrapolation of the rates of the vital components of population growth but in calculating the resulting future population size as well.

3.9 | Reflection

The first signs of an imminent shift in standard population forecasting methodology became apparent at the turn of the century. In the Netherlands these signs became apparent in the nature of the reception and acceptance of the new forecasting methodology by the leading statisticians. Malthusian (geometrical) forecasting methodology of total population lost ground to demographic (component) forecasting methodology. Belief in the existence of a law of population crumbled. It was replaced by long term demographic forecasts in which theories of future population rooted in empirical observation, in ever extending time series of the vital components of population growth and in the findings of social economic and demographic research, played a significant part. Because orthodox forecasting methodology had failed them, the early innovators had to come to an understanding with the future of population in assumption making. The necessity to find theoretical answers to the question what the long term future of population and the levels of the vital components of population growth were going to be was at the root of what made them the earliest contributors to demographic transition theory.

conceptualisation of population dynamics in which the vital components of population growth (Fahlbeck) and age (-sex) structure related sets of risks of exposure (Cannan and Westergaard) played an important part. Leading the impending innovation of forecasting methodology was more than a mere change of methodology, for it had to be preceded by a changed concept of demographic reality. It took on the proportions of a paradigm *change*. Certainty was replaced by speculation on the future course of population if not by indications of its general direction. The concept of societal processes

Figure 3.4. Methorst's diagram. Birth and death rates of the Netherlands, 1876-2009.
Source: Methorst, 1922

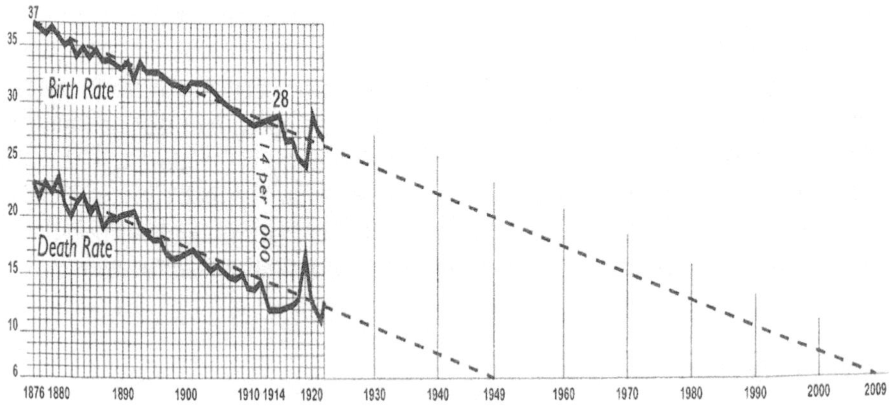

being governed by laws was replaced by social economic and demographic analyses and extrapolations. However, the indications of a Kuhnian paradigm *shift* in stead of a paradigm change are few and vulnerable to poly-interpretation. If a paradigm shift took place, it was not caused by the new, demographic statisticians such as Fahlbeck, Westergaard, Verrijn Stuart and Methorst and the economist Cannan were well informed of the state of the art of demographic analysis. It was the new speculative approach to the future that posed the biggest problems, affecting the statisticians among them in particular.

From then on the community of statisticians had to face three important questions: Was demographic forecasting as a form of speculation about the future course of population to be taken seriously? Was it to be taken as a serious activity of statisticians? Did it belong to the domain of statistics?

The third question was particularly relevant to Dutch statisticians such as Verrijn Stuart and Van Zanten. These men saw statistics as public statistics in the first place. Statisticians were the guardians of the reliability of public statistics. Mathematical statistics, probability calculus and speculation about the future were beyond the scope of these statisticians. It was the main obstacle to an easy propagation of the new forecasting methodology in the Netherlands until the end of the first world war. In terms of the modernisation of forecasting most of the endeavours of early forecasters appear to have passed the Netherlands by.

The basic ingredients of modern forecasting methodology, life table methodology in particular, were at the disposal of Dutch statisticians, economists and demographers in no less degree than elsewhere. For the time being these ingredients did not induce them to contribute to the methodological innovation by themselves; neither did the example set by Cannan, Fahlbeck and Westergaard. It may be assumed that the example set by Fahlbeck influenced Methorst when he made his 1922 forecast. It may be assumed as well that Methorst's example added to the propagation of demographic forecasting in the Netherlands in the inter-war period, because of his position in Dutch statistical circles. It is even possible that Wiebols was familiar with Westergaards' 'horoscope' through Verrijn Stuart's manual of statistics of 1910. However, the inspiration for the true innovation of forecasting, as brought about by Wiebols, came from a different direction, outside mainstream statistics, namely that of mathematical statistics, embodied in the Netherlands in the person of Holwerda.

4. The International Struggle for Paradigm Dominance

4.1 | The Fall of Fertility and the Consequences for Future Net Reproduction

Modern population forecasting took off in the 1920s and 1930s. In these decades, national forecasting was closely related to discussions of the population problem. This problem, however, took different forms in different countries, and in the Netherlands different forms at different times. Except in the Netherlands, population debates in the 1920s were dominated by anxiety about the falling birth rates caused by a decrease in fertility. Only in the 1930s did people in the Netherlands join in concerns about fertility decrease and impending population decline (*Figures 4.1a* and *4.1b*; *Figure 4.2*).

International anxieties were further kindled by Kuczynski's book *The balance of births and deaths* (1928), in which he focused the international attention on the consequences of the ongoing fertility decrease. By using time series of the *net reproduction rate* (NRR) of various countries he succeeded in demonstrating that in most cases fertility had fallen to a level below replacement. He demonstrated that the long standing concept of net reproduction could be used as an instrument with considerable predictive power. A few years before Wiebols had been the first in the Netherlands to point at the insights that could be gained from the application of the concept of net reproduction to forecasting (Wiebols, 1925, pp. 34-42). A decade later Van Zanten (1937) even thought that the use of NRR made population forecasting superfluous as an instrument in discussions of the population problem.

Figure 4.1. Total fertility rates of selected countries of Europe,
a. 1851-1945

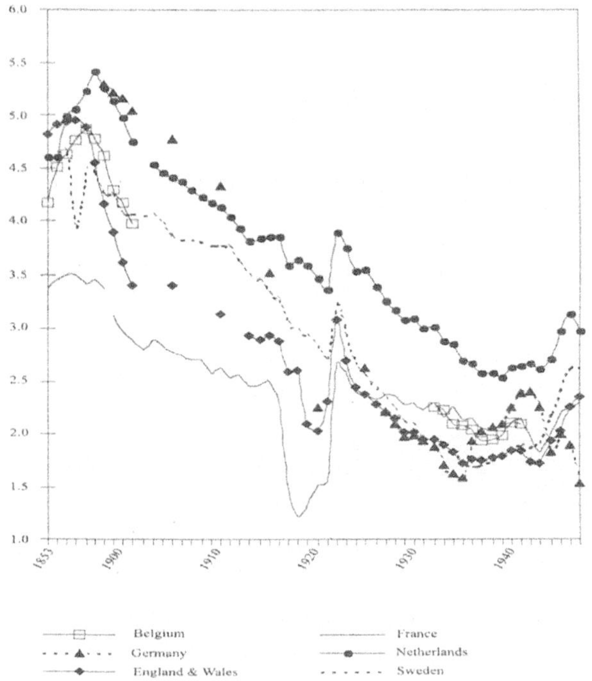

b. 1853-1900

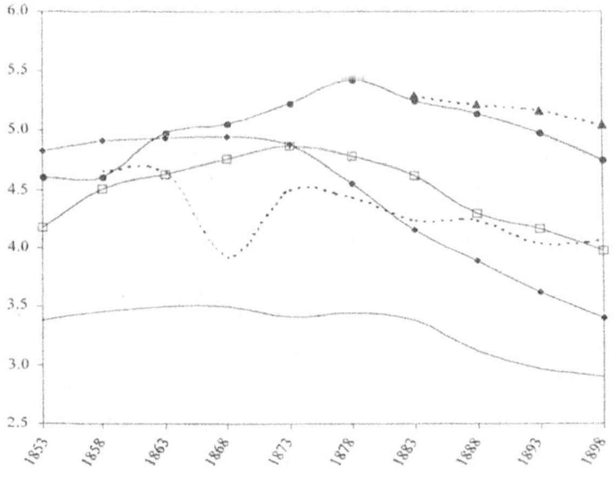

Source: Chesnais, 1992

Figure 4.2. Net reproduction rates of selected countries of Europe, 1930-1939

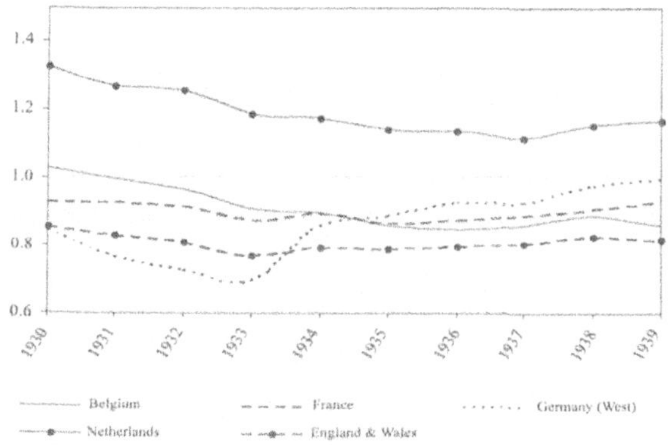

Source: Chesnais, 1992

An attractive aspect of the concept of net reproduction is that it could be readily understood. Nearly all the publications about demographic prophecy in the 1930s, were based to some extent on the 'Kuczynski rate' (Soloway, 1990, p. 234). Due to Kuczynski's work, no-one could any longer deny the facts regarding the feeble reproductive power of European populations. It seemed clear that fears of future under-population, even race suicide had solid quantitative foundations in many countries. The one-sided interest in the fall of fertility and its consequences helps to explain why migration was rarely taken into consideration in calculations of future population. The difficulty of passing judgements on the future course of migration, because of a lack of statistical data and the current belief that most European nations were donor countries of migrants in the first place, did not help much either.

The societal aspect of fertility decrease accounts for the interest in demographic component forecasting. Because of the impact of the age structure on future numbers of births and deaths, it also accounts for the growing interest in demographic (cohort survival/cohort component) forecasting. But that is not to say that belief in a law of population, although hit hard by Cannan, Fahlbeck, Westergaard and other contributors to the prehistory of the concept of demographic transition, disappeared from forecasting methodology. It is hard to imagine that it evaporated without trace, even though there are indications of the opposite. For instance, if one only considers the course of development in a country like the Netherlands.

4.2 | The Rising Line of Methodological Improvement in the Netherlands

The innovation of population forecasting methodology in the Netherlands in the 1920s and 1930s derived from interest in the consequences of population growth. The Netherlands was one of the fastest growing countries of that period. In terms of the slackening of population growth, the country lagged behind other Western European countries. It is a demonstration of the difference in pace of demographic transition pictured so vividly by Cannan, Fahlbeck and particularly Westergaard. In the early 1920s a debate on the population problem started for the first time in the Netherlands. It was triggered by Verrijn Stuart who was worried by the continuing growth of the Dutch population while economic development was expected to stay at a stationary or even declining level. In the debate, the outcomes of Verrijn Stuart's orthodox geometrical growth calculations played an important part (Verrijn Stuart, 1921). In reaction to these calculations Methorst had published his vital component forecast (see Chapter 3, Figure 3.4). While Verrijn Stuart intended to demonstrate that continuation of population growth at the level observed in the past decade was impossible and had to stop, Methorst sought to convince the participants of an imminent decrease in population growth (De Gans, 1995a).

One should not be surprised that actuarial mathematics stimulated the break in forecasting methods, given the long actuarial tradition in the Netherlands. People in the life insurance branch were used to working with various 'probability systems'. Life table methodology and the concepts of life table population ('stable population') and zero population growth in life table populations were familiar. Insurance companies were active since the end of the 18th century. Even though these companies had a wealth of experience and statistical information available, if a good insight into the course of mortality of the Dutch population were needed, one had to consult public statistics on the whole population. In 1923 a survey of all existing population census based life tables was published (Van Haaften, 1923). One of the most promising outcomes of the population debate triggered by Verrijn Stuart, was the suggestion of the actuary Holwerda, that a true understanding of the decreasing birth rate and its consequences for future population growth required sets of 'occurrence-exposure rates', in present day terminology, instead of the crude birth and death rates (Chapter 2; also Chapter 5). Only when the parameters of the occurrence-exposure rates are established can one begin to discuss their future changes. We have seen that it was Holwerda who induced Wiebols to write his thesis on the future size of the population of the Netherlands published in 1925. Wiebols's forecast was not the only result of the debate. There were several other endeavours of demographic forecasting, notably that of Oly in 1924 and of 't Hooft in 1927.

Joh. C. Oly was, like Holwerda, a mathematical advisor to an insurance company. Like Methorst, Oly presumed that a decrease in death rates had already reached its peak. More specifically than Methorst, Oly linked together the future number of births and the future number of women in the fertile age bracket, albeit only verbally: Despite decreasing fertility, the absolute number of births could, for the time being, continue to increase as a result of the 'remarkable youth' of the population. Oly reasoned that, because of the low average age of the population, the group of women in the fertile age bracket would increase quite considerably. As long as this group of 'fertile women' had not yet reached its maximum size, the absolute number of births could continue to increase, despite the decreasing fertility rate (Oly, 1924). The current debate induced him to use his knowledge of life table populations and stationary population theory to put forward his own view of the maximum future population size to be expected.[43] His contribution is a lesson in the application of stationary population theory in forecasting practice. The results of his calculations were presented at the March meeting of the Society of Insurance Science and published in the same year (Oly, 1924). His main interests were the ultimate total population and age-sex structures resulting from two alternative futures. The first of these was the future that would result from a stationary population with life table mortality rates (level 1920) and constant numbers of births (level 1920). In such a theoretical stationary population, both the crude death and birth rates were 18 per 1000 of the population; the ultimate stationary population would amount to about 11.3 million inhabitants in 2020 (Oly, 1924; 1930). The second, more realistic future started from similar mortality rates and a gradual decrease in the crude

[43] His paper contains an interesting age-period-cohort approach and stereogram of the observed course of age-specific mortality probabilities (ten-year age groups) based on Dutch life tables of the period 1840/1851 to 1910/1920.

birth rate from the observed figure of 26 per 1000 in 1923 to 18 per 1000 in 40 years (in 1963).

Oly made a point of stressing that the future number of women in the child-bearing age groups would continue to increase, because of the increase of the number of births in the previous period. He calculated that the decrease of the general fertility rate of women aged 18-43 years still had a long way to go (from 69 per thousand in 1923 to 46 per thousand) to assure the crude birth and death rates reached the necessary level of 18 per thousand each in the ultimate stationary situation. In this future, total population would increase to a total of 14 or 15 million at least, compared with a total of 11 million in the first future. Oly's prudence in answering questions about the degree of overpopulation in the Netherlands is striking, although he foresaw some problems, including the ageing of the population. How the problems were to be solved he could not say: '*We may consider ourselves to be happy, if we can learn something about the future in virtue of the past*' (Oly, 1924, p. 166).

One year later, but presumably independently of Oly, Wiebols' forecast of the future size of the population of the Netherlands was published. Wiebols doubted whether Methorst's diagram with linear decreasing graphs of the birth and death rate would ever become true (Wiebols, 1925, p. 30 f.n. 2). On the other hand he considered it Methorst's great merit that he had focused attention on the 'disturbing' influence of the age structure on the crude birth rate and death rate, although Methorst had failed to draw the right conclusion. The next logical step would have been to conclude that crude birth rates and death rates were unsuitable instruments for his aims. Data on fertility and mortality probabilities only would allow for a deeper penetration into the matter of the future development of population (op. cit., p. 31).

As did Methorst and the host of other forecasters, Wiebols excluded migration from his calculations. Furthermore, he limited his calculations to the female population. In contrast with Methorst, he thought it crucial to use the age structure factor, and consequently future age-specific fertility rates and age-specific probabilities of death (op. cit., p. 31 et seq.).[44] Whereas Methorst and Oly had found verbal emphasis sufficient for the importance of the age

[44] Wiebols used a ten-yearly (general) fertility rate: The number of children born in a decade divided by the number of women between the ages of 15 and 49 who were present at the beginning of the decade (Wiebols, 1925, p. 57).

structure, it became the basis of Wiebols' calculations. Much to his dissatisfaction, he had to put up with the general fertility rate in stead of a set of age-specific fertility rates because of a lack of statistical data. In making it such, Wiebols built on the views of Böckh and Böckh's former student Rahts regarding the population's level of net reproduction and the necessity of using age-specific fertility rates (op. cit., pp. 38-40).

Wiebols presumed that the decrease in the general fertility rate in past decades, down to the level of 57.59 per cent in the period 1909-1920, would continue as an asymptotic curve until in the year 2000 the Netherlands would reach the level of France in the period 1910-1911 (36.59 per cent). In those days this was the lowest known level in Europe (Wiebols, 1925, pp. 59-60). He made a similar assumption for the death rate: The existing death rates would develop asymptotically until in the year 2000 mortality would be the same as in a mortality table of the '5th civil servants table' (1891-1914) but now with ages x+2 instead of x. The death rate within the civil service was assumed to be the lowest observed within the Netherlands (Wiebols, 1925, p. 73).

Wiebols' forecast is a true demographic forecast, starting from age-sex structure (females only), age-sex probabilities of survival derived from life tables, and general fertility rates. Wiebols introduced the concept of Net Reproduction (and the method of its calculation), to demonstrate that using age-specific fertility rates in population forecasting was even better. Lastly he showed, albeit with a theoretical model, how to apply demographic forecasting methodology at the municipal level (Chapter 2).

In 1927, in the aftermath of the first stage of the national debate, another addition augmented the Dutch collection of demographic forecasting approaches. In that year 't Hooft, an engineering mechanic with an interest in the population issue, made a population forecast only slightly different but presumably independent of Cannan (Chapter 5, Figure 5.2). Like Cannan's forecast it was based on ten year cohorts whose numerical history was reconstructed from censuses held at ten year time intervals over the period 1830-1930 and extrapolated by 10 x 10 'still present in the Netherlands'-proportions.[45] The numbers of live births in future ten year time intervals were

45 Like Cannan, 't Hooft was convinced that no immigration surplus would occur. In fact his proportions 'still present in the Netherlands' were equal to proportions 'surviving in the Netherlands'.

calculated on the assumption of a further decrease in numbers until natural growth became zero (De Gans, 1995a).

Within five years national population forecasting in the Netherlands had leapt from geometrical extrapolation to demographic forecasting while the tool kit of forecasting methods was fitted with three different demographic forecasting 'models': Those of Methorst (1922; see Chapter 3), Oly and Wiebols (1924 and 1925 respectively) and 't Hooft (1927a; 1927b).

4.3 | The Dutch Pioneer Position Assessed

What was new in the forecast of Wiebols in comparison to the international pioneers of the preceding period, Cannan, Fahlbeck and Westergaard, and in how far were the forecasting endeavours of Wiebols and the other Dutch innovators of population forecasting methodology ahead of those of their contemporaries of the 1920s?

The predictions of Cannan, Fahlbeck and Westergaard and Wiebols have in common that they are all characteristically long-term predictions, with horizons of sixty years (Fahlbeck), eighty years (Wiebols) and hundred years (Cannan, Westergaard).

The quantitative forecasts of Cannan, Fahlbeck and Wiebols have in common the exclusion of international migration from the calculations. Cannan and Wiebols excluded migration because they could not imagine their respective countries being subject to an immigration surplus. As they saw it, by omitting international migration they merely neglected a possible future emigration surplus. Per definition, their forecasts were therefore maximum forecasts. This suited their purposes well, because they were primarily interested in the indication of the maximum size of the population. Fahlbeck refrained from taking international migration into account because he was only interested in the future course of birth and death rates. His forecast is limited to the extrapolation of the observed trends of the rates of these vital components of population growth.

The innovative aspects of Fahlbeck's contribution were in the domain of population theory, because he refused to assume an ultimate stationary situation; and in assumption making, because of the integration of the findings of social economic theory with fertility decrease. The novelty of Westergaard's qualitative scenario was in its imaginative, but realistic, picture of the demographic transition and its societal implications.

Wiebols' forecast, which can best be compared to that of Cannan, was innovative, because of its completeness, sophistication and application of the state-of-the-art methodology of demographic analysis. Wiebols' forecast was based on a dynamic approach, while Cannan's method was based on a 'comparative state' approach. Cannan used age-specific 'survival in England' proportions, derived from a comparison of the population of successive age groups in successive censuses, instead of true survival rates. Cannan's method is unsuitable for urban and regional forecasting, while age-specific marital fertility and migration could easily be inserted in Wiebols' model, as he demonstrated himself theoretically (respectively Wiebols, 1925, pp. 110-127). Cannan's treatment of future fertility, though pragmatic and founded on sound demographic reasoning, is too simple from the point of view of forecasting method. This is not to diminish the innovative power of Cannan's forecast, his pragmatic and ingenious use of census data, his diagram and, above all, his cohort approach. Although, for instance Kersseboom had already pictured the methodological outline of demographic forecasting in the 18th century, Cannan can be considered to be the first of the modern era.

The international pioneer position of Dutch population forecasting in the first part of the 1920s can best be assessed by a comparison of the forecasts of Oly and Bowley. For, as we have seen, Bowley is generally considered at the origin of the true take off of demographic forecasting in the 1920s. In the same year in which Oly presented the outcome of his calculations, English economist/statistician Bowley's population forecast of Great Britain was published in *The Economic Journal* of June 1924 (Bowley, 1924). In intellectual histories of population forecasting Bowley's forecast is referred to as one of the first forecasts of a population by sex and age after World War I (see Chapter 1; also Wolfe, 1929; Keyfitz, 1972; United Nations, 1973; Esenwein-Rothe, 1982; Pflaumer, 1988). A comparison of the forecasts of Oly and Bowley forms an excellent point of departure for an assessment of the innovative quality of early Dutch forecasts of the formative period.[46]

---

46 Bowley was not the first in England after the First World War to apply probabilities of survival to an existing population. Brownlee had preceded him in 1922. He used the same *English Life Table No. 8* as Bowley in 1924 (Greenwood, 1925). In 1925 Bowley's traces

Bowley was a student of Cannan at the London School of Economy and Political Science (LSE); in later years they became colleagues when Bowley was also appointed to LSE. As a statistician, Bowley was well known in the Netherlands. His book *Elements of statistics* was frequently referred to, for example by Verrijn Stuart (1928) and Van Zanten (1927). As a student Bowley had been inspired by Cannan's forecast. In his own lectures, Bowley made frequent use of Cannan's diagram (De Gans, 1994b). After about thirty years he was the first to build on the foundations Cannan laid. The context of his forecast, its aims and his interpretation of the results of his calculations become clear from a summary made during a debate among English statisticians on forecasting methodology in 1925. He intended "*.. to work out the future age-distribution and total of population on certain clearly defined hypotheses, namely, constancy in number of births and unchanged death-rates. I nowhere imply that I anticipate this constancy. I begin by saying that it is interesting to inquire what birthrate is necessary to prevent a decrease of population, and what would be the ultimate age-distribution in a population in which the number of births was constant and the death-rates stationary*" (Discussion, 1925, pp. 80-81). His intentions were obviously to apply stationary population theory to the societal discussion on the consequences of the observed decrease of British fertility.

The settings of the calculations of Bowley and Oly were completely different. The context of Bowley's forecast was the continuing fall of fertility in his country. Bowley wanted to demonstrate that the fall of the birth rate of Great Britain from 1876 and the decrease in the absolute number of births per annum from 1903 had far reaching consequences for the age composition of the population, specifically for the numbers of people of working age. His main conclusion was that "*at the birthrate (per potential mothers) of the years 1921-23 the population will ultimately diminish if there is any emigration, unless the death-rate falls further*" (Bowley, 1924, p. 192). Because of an observed emigration surplus in previous years, he expected the population of Great Britain to diminish from the year 1941 onwards (Bowley, 1924).[47] In doing so he added fuel to the fear in England of the menace of future under-population (Soloway, 1990).

were followed by Greenwood.

47 The 'emigration surplus' was not included in his calculations.

The demographic situation in the Netherlands was different. Although the birth rate was also decreasing, it was overpopulation, not under-population that was feared. Neither the economic consequences of the World War, nor a short-lived sharp rise in the mortality figures caused by an influenza epidemic, nor the declining birth rate seemed to be capable of checking population growth. In 1921, population growth was higher than ever, while the economic situation –after a post-war economic boom– was worsening again. Unemployment was rising and national poverty and distress seemed to be imminent. Oly was therefore particularly interested in the future upper limit of total population. As an insurance mathematician he was familiar with life table methodology and was able to contribute to the debate.

The forecasts of Oly and Bowley, published in the same year, are strikingly similar. In terms of modelling there are hardly any differences. Both started from stable (stationary) population theory. In both forecasts life table probabilities of survival were used, which was an innovation in comparison with Cannan's forecast, as was the distinction drawn between the male and female parts of the population. With respect to fertility both started from constant future numbers of births. Oly however calculated a maximum variant in which he used dynamic birth rates (a decrease from 26 to 18 per 1000 in 40 years); his scope was much wider. While Bowley's forecast was a conditional forecast, that of Oly came closer to a true forecast. Despite his interest in the upper limit of future population, he tried to achieve insight into the range of the future population size. In doing so he was the first to calculate alternative futures of the Dutch population.

Like Cannan, both Oly and Bowley were aware of the impact of changes of the population composition, particularly the proportion of women in the fertile age category. While Cannan used the concept of general fertility, both Oly and Bowley introduced the actual observed general fertility rates to demonstrate the plausibility of the assumptions made with respect to the future number of births. General fertility rates were not yet integrated into the actual forecasting model. This was left to Wiebols, one year later.[48]

[48] For that reason Keyfitz (1972, p. 349) is mistaken in assuming that Bowley "*followed each sex down its cohort line for survivorship, but used women 20 to 44 years of age as the base for births*". Bowley used constant numbers of births.

In terms of methodology and sophistication, it can be concluded that the early Dutch forecasters can stand comparison both with their predecessors and their contemporaries and were definitely forerunners in the introduction of demographic forecasting.

4.4 | The International Propagation of Demographic Forecasting

The Netherlands and Great Britain were not the only countries where the demographic approach to population forecasting was gaining a foothold. Within a few years, the method of forecasting future population by sex and age was flourishing (Wolfe 1929; Spengler, 1936; United Nations, 1973). As early as 1922 Strumilin had initiated the first demographic population forecast in the Soviet Union. It was worked out by Tarasov at Gosplan (planning horizon 1920-1942) with a twofold motive: To estimate the human losses of the First World War; to forecast the consequences for the labour force. It was based on the population of the All-Union census of 1920, on a life table made by Novoselski in 1916 from data from the 1897 census of the European part of the Russian Empire and on a constant ratio of the number of life births and women in their reproductive ages.[49]

In Sweden, the mathematical statistician Harald Cramér made formal demographic forecasts of the future age structure of the population in 1925 (Cramér, 1935), as did S.D. Wicksell in 1926 (Cramér, op. cit.; Jensen, 1931). Lotka made a formal population forecast of the United States in 1925 (Gini, 1931a). In 1928 P.K. Whelpton made a demographic components forecast of the United States (Keyfitz, 1972; Smith and Keyfitz, 1977; Hecht, 1980). In Italy Corrado Gini made the first demographic forecast in 1926 on the occasion of his inauguration as president of the *Instituto Centrale di Statistica del Regno d'Italia* (Gini, 1931a), followed by Felici Vinci in 1927 (Hecht, 1980).

---

[49] The author is indebted to Dr A.G. Volkov, Head of the Department of Demography of Goskomstat in Moscow, for sending copies of articles (in Russian) on the history and methodology of population forecasts in the USSR and of articles by leading demographers during the 1920s and 1930s on the population forecasts they made. The author is indebted also to Peter Boelens, researcher at the Population Research Center of Groningen University. Boelens made these texts available by writing a paper on population forecasting in Russia, based on these copies, as part of an assignment in the author's PDOD course on *Population and Household Forecasting* (Boelens, 1994).

In France Sauvy made the first demographic forecasts in 1928 and 1929. For a long period of time his forecasts were isolated incidents. It was only after the Second World War, when population growth in the Third World created anxiety in western industrialised countries, that French demographers again became interested in demographic forecasting (Lachiver, 1987).

In Germany it was a different matter. The first demographic forecasts were made by the National Statistical Office of Germany (*Statistisches Reichsamt*) in 1926 (Pflaumer, 1988). No demographic forecasts before that year are known; presumably there was no need for such calculations while the age pyramid and the growth rate gave no grounds for anxiety (Höhn, 1986). Demographic forecasts were published in 1926, 1930, 1938 and 1939; Bürgdörfer, Benser and Hage respectively were responsible (Esenwein-Rothe, 1982; Höhn, 1986). The 1928 forecast of the *Statistisches Reichsamt* is an impressive piece of work, both for its methodology and the presentation of data. It built on the 1926 forecast, but the calculations had a better foundation from a methodological point of view. Three variants were calculated, based on the assumption of a constant mortality regime (the life table 1924-1926) and three fertility assumptions: Total number of life births constant at the level of 1927; a further decrease of marital and extramarital fertility between 1927 and 1955 of 25 per cent; constant marital fertility (over a part of the forecasting period only) as was usual in almost all western European forecasts between the two world wars. Migration was also left out of the German calculations; that Germany would see a structural immigration surplus was unimaginable. Specific attention was paid to the consequences of the forecast for the age structure, the productive and the unproductive parts of the population and to the consequences for disablement insurance. Lastly, forecasts were made of the populations of other European countries (England and Scotland, Italy, Netherlands, Sweden, Norway, Denmark, Poland and Ukraine) based on the most recent life tables and the assumption of constancy in the total number of live births (Statistisches Reichsamt, 1928).

4.5 | The Struggle for Paradigm Dominance (Geneva 1927)

If one only considers the course of developments as pictured in the preceding sections, one would answer the question whether belief in a law governing population growth quietly disappeared positively. A more thorough analysis, however, reveals that the actual course of events was completely different.

The belief in a law of population got new and strong impulses in the early 1920s and even seemed to (re)gain dominance for a short while.

As time passed and the decrease of the birth rates of many countries continued, population forecasting methodology became a discussion topic at international conferences in the context of the debate on the presumed future population problem caused by the fall of fertility. International conferences particularly vital to the propagation of knowledge of methodological innovations in the inter-war period were the World Population Conference of Geneva in 1927 and the Session of the International Statistical Institute in Tokyo (1930). In Geneva the focus of attention was on the merits of a new law of population growth (the logistic law). In Tokyo the issue was the decrease of the birth rate and its consequences for future population growth; here, oral and written contributions almost exclusively reflected the new demographic approach to forecasting methodology. Future population size and demographic forecasting methodology were discussed at the conferences of the International Union for the Scientific Investigation of Population Problems in London (1931) and the International Congress for Studies Regarding Population in Rome in the same year. At this latter conference, governments were recommended to make demographic forecasts in imitation of those of the German Statistisches Reichsamt (Paulinus, 1935/36). In 1931 (London) attention was equally divided between the demographic and the logistic approaches; in Rome the focus was on demographic forecasting.

At the 1927 World Population Conference, sociologists and biologists met for the first time "*in the solution of economic problems*" (Sanger, 1927, p. 13). The conference witnessed the collision of two methodologies; the logistic and the demographic. It was a repetition of the debate between the statisticians Yule, an advocate of the logistic method, and Bowley at a meeting of the British Royal Statistical Society in 1925 (see Yule, 1925; and Discussion, 1925). At the Geneva conference of 1927 the merits of demographic forecasting seemed to fade under the curve-fitting power of the recently rediscovered mathematical law of logistic population growth. From a present day point of view this is surprising, even if it is remembered that in most countries students of the population problem were primarily interested in the increase of total population rather than in future age-sex structures. Moreover, in the opinion of the propagators of the logistic method of population growth, the method had proved to be a good fit of the observed course of populations.[50]

Finally it must be borne in mind that the American biologist Raymond Pearl, a (re)inventor of the logistic growth theory, was one of the key scientists at the World Population Conference of 1927.[51] As a biologist Pearl was inclined to look for similarities of growth in differing populations. He was convinced that population growth, with respect to its rate, appears to be a fundamental phenomenon in which insects and mankind behave in a similar manner (Hiller, 1930). In 1928 Pearl was elected the first president of the International Union for the Scientific Investigation of Population Problems set up as the permanent successor to the World Population Conference (Pitt-Rivers, 1932).

The impact of the new curve-fitting methodology on population forecasting was enormous. According to Wolfe (1928), no contribution to the theory of population for decades had commanded the attention which Pearl's logistic formula received. Explanations of its popularity were Pearl's scientific prominence as a biometrician; the new interest of biologists in human population problems; the fact that the new 'law' of population growth seemed to free the world from the old Malthusian fear of overpopulation. Wolfe's last mentioned argument was most vividly expressed by the mathematician-statistician G. Udny Yule at the 1924-25 session of the Royal Statistical Society: The outcome of the work of Pearl, Reed and their colleagues was to portray the growth of a population as a biological self-regulating process, "*indeed, a process of which the regulation is extraordinarily sensitive*" (Yule, 1925, p. 40). The logistic formula also became popular because as a mathemat-

50 The advocates of the logistic growth approach had good certificates. The method gave, for instance, a good fit of the population of England and Wales over the period 1901-1911 (Yule, 1925). In the thirties Alfred J. Lotka (1932) found a similar goodness of fit for the population of the United States.

51 The other one was Lowell J. Reed from John Hopkins University. The mathematical equation of logistic growth had proved to be useful as a descriptor of observed growth of a wide variety of organisms, including human populations. The logistic growth curve is "*.. S-shaped, concave before a point of inflexion, thereafter convex, and gradually approaching (but never reaching), at a continuously retarded rate, an upper limit or asymptote*" (Wolfe, 1928, p. 679). The formula was discovered by the Belgian mathematician Verhulst as early as 1838. It was used in the first place for curve fitting purposes. A simple extension of the procedure provided the desired projections. Like all mathematical curves, the logistic is quite mechanistic; it cannot be used for the projection of decreasing populations (Shryock and Siegel, 1973, p. 177).

ical expression it linked the study of population with 'true' science. As Wolfe put it, the formula was in line with the ".. *scientific fashion, that is the extraordinary postwar development of mathematical statistics and the attempt to apply statistical methods to all sorts of problems, not always with adequate attention to the question whether they are amenable to valid methods of statistical analysis*" (Wolfe, op. cit.).

Apparently the dream of laws determining the growth of population was still vivid, limiting human intervention to the task of finding these laws and the parameters of the mathematical functions in which these laws were supposed to be expressed. The fascination was expressed most clearly by Thomas Adams, originator of the first Regional Plan of New York and Environs (1928) and one of the first to introduce logistic growth methodology into regional surveying: " *The contour of the extended or predicted portion of the resulting curve is fully determined by the past population figures, rather than by the judgment of the estimator. This very greatly reduces the play of personal judgment in making the predictions.*" (Adams *et al.*, 1929, p. 110). At the same time, Adams was well aware of its temptations, for he continued: "*The result has a dangerous fascination because the prediction seems to be growing out of the past and to be an inevitable statement of what the past implies for the future. Actually, however, the personal element is by no means absent, for in the preliminary work of fitting the equation to the observed data judgment and opinion play a considerable part.*" (Adams *et al.*, op. cit.).

A political economist himself, Wolfe blamed statisticians, whether in the natural or social sciences, for reasoning too readily from a frame of mind in which empirical formulae are tacitly regarded as laws that govern societal phenomena. For in doing so, they added a *'mystical virtue'* to the mathematical equation (Wolfe, 1928, p. 680).

The logistic method was also criticised from a social-cultural point of view. Hiller (1930), for instance, pointed out that in the end, a *general* population theory as the logistic should rest on cultural premises. The belief that such a theory could imply a single or uniform factor as the cause for the varying rates of growth had to be excluded. Such a theory would search for distinctive types of causal factors in different social groups, and would assume that varied combinations of material and non material cultural elements might account for observed trends in statistical data. Such trends measure, but do not disclose or explain the actual processes involved. They are normally mediated through

Vignette 7: The World Population Conference of 1927

The initiative for the organization of the conference was taken by a group of biologists '.. a small group of American scientists headed by East, Little and Pearl, all well known in eugenic circles. The ideas of these men were further developed by English scientists with at the head Carr-Saunders, Crew and Huxley. Mrs Sanger was acting as its secretary and it is for the greater part her merit that, as early as 1927, a World Population Conference could be organized in Geneva' (Ons Nageslacht). Margaret Sanger was the driving force behind the organisation of the conference and also one of the initiators of the formation of the Population Association of America. She was a 'politically astute' advocate of birth control and of women's right to determine whether she will bear children or not and how many. She was the instigating force behind the National Birth Control Conference of New York in 1921 and the Sixth International Birth Control and New-Malthusian Conference of New York in 1925. Her successful arrangements of conferences was due to her easy access to funds through the many society contacts she had due to the wealth of her husband, and to her great organizing ability (Hodgson, 1991). In a Dutch report of the proceedings of the conference its character and aims were summarised as follows: 'To start an international discourse on the population problem and on population science in which biologists, physicians, sociologists, economists, statisticians and politicians would participate. The call to this conference was the population problem, which was sharply felt by various nations; each nation being as a rule well aware of the population problem of other nations. The nations do not fail to see that the increase of population forms a menace to the future. Notwithstanding the world population problem is one of the few problems that is not at the agenda of international co-operation' (Ons Nageslacht, 1929). The interest of American biologists in population studies was closely related to eugenics. Their central concern was the biological consequences of population trends. Other than Pearl's logistic theory of population growth, biologists contributed little to the study of the determinants of population trends.

social factors subject to change. Therefore, any attempt to predict long-term population cycles would probably go as far astray as did the forecasts of Malthus. Similarily Bouthoul (1935) thought that many advocates of the 'law' belief started from an interpretation that was too simple (see Chapter 1).

It is small wonder that so much attention was given to the logistic method at a conference where Pearl took such a prominent part. Pearl presented the opening paper of the conference, on *The Biology of Population Growth*, in which the merits of the logistic method as a curve fitting method was central. From the debate following his presentation, it can be learned that Pearl did not overestimate the qualities of the logistic 'law'. Although he had found that populations closely followed the logistic curve so far as their historically recorded past history went, he admitted that the observation did not imply that populations would continue to follow the same particular logistics for the indefinite future. It was not his intention to make "*so rash a prophecy*". All he ever said was that, if the same forces, biological, social, economic or any other, which have in the past caused these populations to follow a particular logistic curve continue to operate in the future unchanged, it seemed a not unreasonable inference that the populations would continue to follow this same curve in the future (Sanger, 1927, p. 55). Four years later, at the second General Assembly of the International Union for the Scientific Investigation of Population Problems of London, Lotka presented a formal demographic model to which a population developing under the regime of such a logistic law of population would conform. He set out what would be the typical trend of the birth rate, the net fertility, the death rate, the distribution by age and sex, the discrepancy between the 'true' or 'inherent' rate of natural increase in such a population. In doing so he applied his findings in the field of formal stable population theory, to logistic population theory (Lotka, 1932).

One would expect the demographic approach to the calculation of the future of population to find advocates at the World Population Conference. It is remarkable, therefore, that in the debate of the merits of the logistic law following the presentation of Pearl's paper in 1927, Bowley's demographic approach was advocated by one discussant only, R.A. Fisher, an economist and statistician. Fisher admitted that the logistic equation provided an admirable curve for interpolation of total population over short ranges "*provided we have not any other important relevant information*" (Fisher, in: Sanger, 1927, p. 45). But, so continued Fisher, contrary to the method of Bowley, the logistic did not allow for a future decrease of population after a period of increase. In population censuses, all relevant information on the age structure of human populations was available, which was in his opinion of even greater significance than the census totals. The census data allowed for much more accurate predictions than those that could be made from any curve of census totals. It would be foolish to ignore the existence of such data. On the basis of these data, one would arrive at conclusions very different from those indicated by the logistic curve (Fisher, in Sanger, 1927).

Fisher's point of view was not publicly supported by any of the other conference participants. But because he took part in the debate, it is likely that his intervention contributed to the international reputation of Bowley's forecast, whether he intended that or not.

In an indirect, non provocative and presumably unnoticed way, however, there was an intervention from Methorst that could be interpreted as an implicit form of support. In another conference session, Methorst presented a paper on differential birth rates in the Netherlands. Methorst was definitely not well equipped for a debate on mathematical matters. For that reason it was presumably sensible for him to refrain from any reference to the debate on the merits of the logistic method compared with the demographic method. In his paper he confined himself to an account of what can be read as a summary of the essentials of his 1922 forecast: "*The temporary decline of the death rate as the result of hygiene causes the increase of population; the sense of responsibility promoted by hygiene causes the birth rate to go down. After having gone down to a certain minimum, the death rate goes up again. Thus the two curves of birth and death rate are, as it were, predestined to approach each other again, in accordance with a most regular evolution, and the population problem is solved.*" (Methorst, 1927, p. 179). Methorst's contribution can be read as a prudent support of the demographic approach. Other than to Fisher, Methorst refrained from informing the audience of the methodological advances in his home country. Given the advanced state of forecasting in the Netherlands by that time, one cannot help thinking that his statement on the future curves of the birth and the death rate was shallow. He limited himself to an unaltered repetition of his views of 1922, coloured by notions of demographic parallelism (see Chapter 3). At the time of the First Population Conference, a debate on the best approach to the calculation of the future size of the population was raging in the columns of the Dutch economic journal *De Economist*, with Holwerda and Wiebols as the advocates of the merits of the new demographic method. One has the impression that the debate had passed Methorst by (De Gans, 1995a; also Chapter 5).

4.6 | The Paradigm Change Accomplished (Tokyo, 1930)

In 1930, at the Conference of the International Statistical Institute (ISI) in Tokyo, a whole session was devoted to the consequence of the continuing fall of the birth rate in most Western countries. This time the setting was different. No American biologist such as Pearl set the tone, but the Board of ISI itself. When the Tokyo Session was held, Bowley was the Treasurer and Methorst the Secretary General and Director of the Permanent Office. Both men had distinguished themselves by making demographic forecasts. At the 1925 Meeting of the Royal Statistical Society, Bowley made it clear that he was a fervent advocate of his own demographic approach to forecasting and was opposed to the use of mathematical extrapolations on the basis of logistic growth methodology.

In December 1929, the Board of ISI decided to devote a session to the subject of the future of populations and to invite certain ISI members to write papers on the subject of "*the tendency of population with the present birth and death rates*".[52] The topic busied the minds of statisticians, demographers, and economists in many countries and resulted in a series of studies on the estimation of the future evolution of the population. The declining birth rate was the main demographic issue of the 1930s. Two years later, for instance, the issue stood central at a meeting of the Belgian *Société scientifique de Bruxelles* (1932). The theme of that meeting is of interest because it highlights the existing anxieties even better than that of the Tokyo conference: '*What number of marital births is necessary to maintain the current Belgian population, given a constancy of the other relevant data: Longevity, mortality, age at marriage, divorces, remarriages, et cetera.*' (Vulhopp, 1932, p. 28).

Given the composition of the board of ISI and the theme of the conference, there was not much room for a discussion of the merits of the logistic method

52 Letter of 24th February 1930 from Methorst to Bowley (Archives ISI B.04.03).

Vignette 8: The Sessions of the International Statistical Institute ISI

The Tokyo conference was an extra ISI Session, heavily subsidized by the Japanese Government. The sessions of ISI stand in a long tradition. From 1853 to 1876 nine international statistical conferences were organized. Apart from the statistical conferences there were international demographic conferences later united with hygienic conferences into International conferences on hygiene and demography, bringing together key international statisticians and demographers. On the occasion of the celebration of the 25th anniversary of the Statistical Society of Paris in 1885 it was decided to establish an International Statistical Institut in order to further the advancement of public and scientific statistics. From then on every two years an international conference (Session) was organized. After preliminary consideration of the idea at the 1909 and 1911 Sessions of ISI it was decided in 1913 to establish a Permanent Office, under the authority of the Institute for the purpose of publishing an annual of international statistics, of maintaining a library, of keeping the Institute's archives and of preparing the programs of the Institute's sessions. The Secretary General was to be the Director of the Permanent Office; its seat was located in The Hague. From its foundation in 1885, the collection and publication of international statistical data had been ISI's main task. The Institute exerted a direct influence on the statistical practices of governments. ISI contributed to the innovation of modern demographic forecasting in several ways. By furthering the collection and international comparability of population statistics; contributing to the propagation of information about instruments for demographic analysis and demographic forecasting; facilitating discussions about the expectations of future population and the merits of demographic forecasting. After the Second World War most of ISI's traditional tasks were taken over by other international organisations, among others the World Health Organization. ISI had the features of an aged, decaying institution. There had been a failure to recruit innovative young leaders in statistical development. The revitalisation of ISI began from 1947 onwards. The focus was now turned to studying statistical theories, appraising statistical methods and practices, encouraging statistical research, and furthering the use of statistical methods in diverse subject matter fields wherever useful. The traditional emphasis on official statistics was abandoned; the influence of modern mathematical statistics became dominant (De Gans, 1997, p. 124)

or its application. ISI members, representatives of the statistical offices of various countries, were invited to contribute to the theme.[53] In the event papers were written by Gini, Jensen, Ptoukha and Warren S. Thompson, Head of the American Scripps Foundation for Research in Population Problems. All papers were published in the Bulletin of ISI in 1931 (Gini, 1931a; Jensen, 1931; Ptoukha, 1931; Thompson, 1931a). It is interesting to note that all the contributors of papers are mentioned in intellectual histories of population forecasting. Their papers were listed in contemporary overviews (Carr-Saunders, 1936; Van Zanten, 1938; Glass, 1940) and are referred to in more recent ones (e.g. Hecht, 1980).

The conference proceedings offer an excellent overview of the state of the art of that period.Although direct references to Kuszynski's work are lacking, there is sufficient circumstantial evidence that most authors were familiar with the concept of net reproduction.[54] All papers were based on the demographic approach, although it was evaluated in various ways in comparison with the logistic method. Jensen indiscriminately took the calculations of Bowley, Pearl, Yule and others together (Jensen, 1931). An explicit stand in favour of the demographic method was taken by Gini and Ptoukha.[55] Gini saw a straight line of progress in forecasting methodology: From simple extrapolations of total census populations or calculations based on past observations by extrapolation of a variety of curve fitting functions, in particular the logistic curve of

53 They included Corrado Gini (Italy), Michel Huber (France), Adolph Jensen (Denmark) and Michel Ptoukha (Ukraine). Given the efforts of the voyage and the time it would take, one can understand that many ISI members decided not to participate. Thompson and Gini were able to present their papers in person at the conference, Jensen and Ptoukha were absent. Ptoukha, who was a statistician at the statistical office of the Ukraine, had problems of his own. Presumably because of the political conditions in his country, he was not allowed to be present (Arch. ISI B.04.03). Letters from the Secretary General of ISI (Methorst) to the Directors of the Central Statistical Office of the USSR in Moscow and of the Ukraine in Kharkov, in which the importance of Ptoukha being present in person was stressed, did not help (Arch. ISI B.04.03).

54 Jensen gave the NRR of Denmark of 1929 (0.94320) (Jensen, 1931). Gini had referred to Kuczynski's *The Balance of Births and Deaths* in an earlier paper (Gini, 1930). Thompson (1931) referred not to the NRR but to the 'true' rate of natural growth, which was the counterpart of NRR in formal demography of Kuczynski's rate. The 'true' rate of natural growth is part of stable population theory, worked out by Dublin and Lotka in 1925.

55 Thompson (1931a) referred to demographic studies of the ratios of young children to women of childbearing ages. It was the subject of a book he was working on that was to be published in 1931 (Thompson, 1931b).

Verhulst and Pearl and Reed, via extrapolations of the components of population growth (births, deaths, emigrations and immigrations) to calculations of future population based on the knowledge of age-specific fertility, mortality and migration coefficients (Gini, 1931a). Ptoukha drew a distinction between two kinds of methodological performance. The first one saw population growth as something 'global', relative only to total population, permitting the application of mathematical formulae, the most simple based on the hypothesis of geometrical population growth or –more recently– on 'biological evaluations' (logistic growth). From a purely statistical point of view he preferred the second kind of methodological performance: The estimation of future population by means of a detailed treatment of statistical data on mortality, fertility (and migration) (Ptoukha, 1931).

The discussion gave participants the opportunity to provide supplementary information about the achievements of population forecasters in their respective countries. From the Netherlands Methorst, Verrijn Stuart and Van Zanten together represented contemporary expertise in Dutch population forecasting. Neither Methorst nor Verrijn Stuart took part in the discussions. Methorst merely had an abstract of his views read instead (Séances, 1931, pp. 72-73). The abstract summarised the assumptions of his 1922 forecast and did not differ much from the views he had presented in 1927, with the exception that the abstract was a little more elaborate and departed from former notions close to the theory of demographic parallelism. Methorst stated that population growth was determined by an immigration surplus and a surplus of births over deaths. Of these, only the immigration surplus could be controlled. Mortality can be reduced by hygienic measures; because of the resulting changes in the age structure of the population a future increase in the crude death rate could be expected. The future course of the crude birth rate is much more difficult to calculate. Methorst expected that a further improvement in communications between urban and rural areas and a further abundance of the material and social necessities of life would erase the differences, even affecting the reproductive behaviour of religious families; a decrease in the birth rate in rural areas would arise. He saw the question how this evolution would take place as unresolved, and definitely not one to be calculated in figures. Taking the graphs of birth and death over a sufficiently long period of time to dip into the future would be quite satisfactory for nations with negligible migration movements (Séances, 1931, p. 73). He would need another six years to introduce the age pyramid –as the pivot of the population problem– in his calculation of future population development in an explicit way (Methorst, 1937).

Methorst's digest has the semblance of an *ex cathedra* statement. Of course, he only intended to contribute to the issue under discussion at the conference; the idea of presenting a national state-of-the-art paper on demographic forecasting must have been far from his mind. With the benefit of hindsight it cannot be denied that he missed an opportunity for the international positioning of Dutch population forecasting in a similar way as in 1927.

4.7 | Van Zanten's Intervention on the Statistician's Responsibilities and What it Brought

This left Van Zanten to take an active part in the discussion on the merits of the contributed papers; and so he did. Given the contents of his intervention it is clear that he had a mission of his own to be fulfilled. Van Zanten's long standing and strong conviction that predictions should not be made by statisticians nor statistical offices has been mentioned in the preceding chapters. The task of statisticians was to provide accurate and reliable statistical data. With respect to forecasting they should limit themselves to indications of the best way of making forecasts. He was not against forecasting, as long it was done by students of other sciences, such as economists or town planners.

A passage in Ptoukha's paper may have been the immediate cause of an intervention that changed the direction of the following debate completely. Ptoukha complained in his paper that statistical science, although by nature limited to the provision and treatment of numerical data concerning the past and the present, had already been burdening itself with making provisions in certain domains of the economic and social life (Ptoukha, 1931).

The state of affairs was formulated in a more positive way by Gini. In his opinion statistics, if it were to be taken seriously as a science, could not abstain from establishing regularities and looking for laws, relationships between facts that hold not only for the past and the present but also for the future (Séances, 1931). Gini's view is mirrored in the activities of ISI itself. Ever since its foundation ISI had occupied itself with the question and problems of economic forecasting (Zahn, 1934, pp. 163-165). A long term interest in the future course of the population of its members is seen in reports published on several occasions (e.g. Fahlbeck, at the Session of 1905, and Westergaard at the Session of 1907). As a consequence of the economic turbulence in the period after World War I and the economic crisis at the end of the Twenties the importance of forecasts of the economic situation was increasing again. Recently a committee had been charged by the ISI with the study of the most instructive statistical elements to be collected in the main countries of the world for economic forecasts (Zahn, 1934). The very interest in economic forecasting among the ISI statisticians may even have been one of the main reasons of ISI's focus on demographic forecasting in Tokyo.

Despite this state of affairs Van Zanten thought it suitable to ask ISI to declare that statisticians should refrain from any responsibility regarding calculations of this kind (Séances, 1931, p. 72). Statisticians should limit themselves to establishing what has been (the past) and what is (the present).

Van Zanten's contribution added to the liveliness of the debate, but he offended the officials of ISI who had raised the central theme of the meeting, and some eminent members invited to report on the future course of the population who had just presented the results of their forecasts. It led Gini to write an article in 'his' journal *Metron* that same year in which he reported on the proceedings without failing to include his own part in the debate (Gini, 1931b). The combined information of Gini's article, the official ISI report of the conference (Séances, 1931) and an article written by Methorst (1930) give a clear picture of the issues discussed.[56]

In the course of the debate, it appears that Van Zanten's view was supported by several fellow statisticians, but it did not prevail. A more negative consequence was that sight was lost of the very contents of the contributed papers (Gini, 1931b). On the other hand the debate on the feasibility of population forecasting by statisticians highlighted several other issues thought to be interesting for discussion at the beginning of the 1930s. One of these was the question whether government policies should be based on calculations resulting from similar 'arithmetical gymnastics'. Another concerned the fear that the press would make itself master of the results without taking into account the fact that forecasts were insecure, or without understanding that the observed variation of the outcome of forecasts resulted from dependency on the assump-

56 Methorst, as the Secretary General of ISI, was ultimately responsible for the contents of the report of the proceedings, published in Séances (1931).

tions made. According to Gini, it was remonstrated that forecasts were more than mere 'arithmetical gymnastics': Similar calculations had made clear that the overpopulation danger which was the nightmare of those statisticians who had fixed their attention on the observed surplus of births only, did not exist. Forecasts had brought in the open that most countries in Western, North and Central Europe stood at the eve of the end of population growth. The possibility that the white race might be reduced in numbers in the future was much more than the indifferent outcome of arithmetical gymnastics. Gini judged it to be one of the most significant features of modern statistics (Gini, 1931b). Instead of fearing forecasts, so (according to Gini) it was argued, governments should ask statisticians to make them. Forecasts are indispensable for policy making. If statisticians refused to make them, others would take over, with more mistakes and uncertainties. In several countries national statistical offices even had the duty imposed on them to make population forecasts.

Another issue was the length of forecasting horizons used. Here again, opinions differed. Some statisticians thought the horizons taken were much too long. These criticisms were met by the argument that the horizons were usually not later than 1961, the length of just one generation. Even if the year 2000 were chosen, less than two generations would be involved. Given the length of the existing time series of demographic data, this was considered quite a short period. Moreover the time series used in population forecasting provided a more solid basis for forecasting than in any other domain of forecasting. Some statisticians were even of the opinion that long-term forecasts were more accurate than short term forecasts, because the latter were governed by fortuitous, transitory factors (Gini, 1931b).

Back in Holland Methorst wrote an article in the Dutch economic journal *Economisch-Statistische Berichten* reporting the proceedings of the Tokyo conference. It is striking that scant attention was paid to the future populations topic, but hardly surprising because he had not been present at that meeting. It is interesting, however, to note what items he thought important enough to mention: The names of the countries that had been subject to considerations about their future population growth; a comment on the significance of such considerations, because they gave insight into whether or not the white race would be supplanted by coloured races; lastly, and contrary to Gini's report, that such calculations should be made for short periods of time only, because of changing conditions (Methorst, 1930).

The blunt expressions of the fear of the white race being supplanted by coloured races must have been an interesting experience for the Japanese representatives at the conference. The fear was obviously shared by Methorst and Gini; both mention it in their articles. The debate on the length of the forecasting horizon is not mentioned in the official proceedings of the Tokyo conference (that resorted under the responsibility of Methorst). In comparison with Gini's summary of the discussion, that of Methorst is strikingly partial. Seven years later Methorst saw it as a strong point of his forecast that, although it was still hypothetical, its horizon had been reduced to 40 instead of 100 years (Methorst, 1937).

4.8 | The Importance of Being Present

In the preceding sections the hypothesis of a pioneer position of Dutch population forecasters in the twenties has been confirmed. However, a pioneer position does not necessarily imply that there is a contribution to international discourse on the advancement of the field, neither in an active nor a passive way, nor from being included in the official intellectual history of a scientific field. Now the forerunner position of Dutch population forecasters has been evaluated, its failure to be reflected in intellectual histories of modern population forecasting has to be explained.

A concise example of such an intellectual history of "*population projection with fixed rates taking account of age*" is the one presented by Keyfitz in the seventies: "*This modern approach, based on separate recognition of birth and death components, seems first to have been used by Edwin Cannan* (in 1895 - hdg). *He in effect applied our matrix M, but took a less exact account of the age distribution of childbearing. Bowley* (in 1924 - hdg) *also followed each sex down its cohort line for survivorship, but used women of 20 to 44 years of age as the base for births. The next appearance of the components method, a thoroughgoing exploration that has brought it into very wide use, was by Whelpton* (in 1928 and 1936 - hdg). *It remained for Bernardelli* (in 1941 - hdg), *Lewis* (in 1942 - hdg) *and above all Leslie* (in 1945 and 1949 - hdg) *to show that the procedure could be compactly presented in matrix form, and that such a presentation permits analysis of the regime of mortality and fertility in abstraction from the age distribution by which it is to be multiplied. The continuous form of the same projection was imaginatively studied by Lotka* (in 1907, 1911, 1939 - hdg) *over a 40 year career that created the core of theoretical demography*" (Keyfitz,1972, p. 349; also Wattelar's concise history in Chapter 1). Keyfitz' picture, written from the point of view of a matrix (and a discrete mathematical) approach to forecasting, forms the core of a similar and more recent history in Willekens' overview of the state of the art of demographic forecasting (Chapter 1). In effect the story comes closer to an intellectual history of Anglo-American forecasting than to a general history but, at the same time, it is a good reflection of the dominant position of Anglo-American forecasting in history writing. Moreover, it is difficult to deny the impression of a dominant interest in the mathematical and formal aspects of forecasting in history writing. But this forms only part of the possible explanation. References are also absent in histories of an European origin.

An analysis of contemporary and recent overviews (e.g. Carr-Saunders, 1936, pp. 128-129; Glass, 1940/1967, pp. 468-472 and Hecht, 1980, p. 355 f.n.2) indicates that contributions to international statistical or demographic journals in English, French or German, or to international discourses on developments in the field are likely to be incorporated in intellectual international history.[57] Oly, Wiebols and 't Hooft were active contributors to the national discourse on the method of calculating the future size of populations is discussed in Chapter 5. There is no evidence that either of them ever contributed to an international journal, or participated in an international conference where the subject of population forecasting was discussed. The exchange of information about Dutch achievements had therefore to come from others.

4.9 | Why Dutch Internationals Failed

In order to understand this absence of references it is necessary to look more closely at those occasions which, with hindsight, proved to be vital in the development of demographic forecasting and history writing, namely international conferences and other meetings of students of population. The significance of conferences and meetings of experts and professionals for the exchange of information, the stimulation and inspiration of new developments and the propagation of scientific findings has previously been discussed.

[57] There is of course the bias of dependency on the available written sources. It is not possible to uncover exchanges of information in informal meetings of experts on the occasion of international conferences or in the private sphere.

It should now be clear that the absence of Dutch innovative performances in international overviews of literature, or in intellectual histories of modern population forecasting, is due not to a lack of 'presence' at conferences, nor to a lack of influence in international organisations, but solely because Dutch participants did not think it necessary or profitable to act as intermediaries in the exchange of information on forecasting. Dutch statisticians-demographers held influential positions in international organisations, they were present at international meetings eventful in the history of demographic forecasting and providing excellent opportunities for the exchange of information. Where necessary they kept the national interests well in mind. In the first decade of the 20th century Verrijn Stuart and Methorst succeeded in having The Hague selected as the seat of the Permanent Office of ISI and Methorst invested time and energy in safeguarding the location of The Hague. When the International Union of the Scientific Investigation of the Population was established (in 1928), in line with the resolutions of the Geneva World Population Conference (1927), Methorst immediately organised demographers in the Netherlands into the Dutch branch of the Union and was elected a member of the board of the International Union.

The true innovators of forecasting, Holwerda, Oly, Wiebols and 't Hooft were not in the position to propagate their innovative ideas in international circles. They did not belong to the small group of Dutch representatives at international occasions; Oly, Wiebols and 't Hooft because they did not have the right academic and professional position and did not publish in international journals; Holwerda, who was a visiting professor at the Business school of Rotterdam, because his main interest was in the insurance profession. He was for instance not a member of ISI.

The exchange of information therefore depended on the very small group of Dutch 'internationals' consisting of Verrijn Stuart, Methorst and Van Zanten. Of these, Methorst was the only one who did not abstain from the actual forecasting business, although it is doubtful whether he ever made the necessary calculations himself. In terms of the precise calculation procedures of his forecasts his publications are remarkably unclear. Methorst had strong opinions with respect to the future course of the birth and the death rate and he did not mind exerting his influence in convincing others of his opinions. One gains the impression that he took no part in the debate on methods, either because he lacked the necessary mathematical and statistical background, or because he thought that, as Secretary General of ISI and Director of its Permanent Office, he should refrain from taking an outspoken position.

The Dutch 'internationals' belonged to schools of statistics other than the mathematical school. For that reason they could not be expected to be interested in the advancement and propagation of the formal mathematical modelling part of forecasting. Their view of statistics led both Verrijn Stuart and Van Zanten to oppose the very idea of statisticians making forecasts.

Reading his publications, one gets the impression that Verrijn Stuart's interest lay not in forecasting methodology, but in the future course of the birth rate and how future population growth could be halted. Van Zanten on the other hand was definitely interested in forecasting because he felt it his responsibility as a statistician to provide reliable and accurate public statistics for his clients; he developed into a non-practising expert. In the world of statisticians, however, he obviously had a mission other than the propagation of information on forecasting methodology to fulfil: To prevent statisticians from making forecasts.

4.10 | Growth to Maturity in the 1930s

In the 1930s demographic forecasting of national populations gradually entered maturity. In comparison with the 1920s it was a decade not of methodological innovation, but of general application. In the Netherlands the true methodological innovation of the decade took place in the endeavours to apply the new approach at urban and regional levels, where solutions had to be found for the integration of internal migration in the calculations. Formal mathematical modelling and an explicit search for a standard forecasting model, like those in America by Lewis, Bernardelli and Leslie, seem with only one exception (discussed hereafter and in Chapter 7) to have been outside the scope of interest of Dutch forecasters, statisticians and demographers. Forecasting, at least forecasting at the national level, was still a private endeavour in which each forecaster looked for individual solutions to problems of method.

In countries like Germany demographic forecasting was institutionalised insofar as it became the task of the national statistical office (*Statistisches Reichsamt*). In the Netherlands NCBS refrained from making demographic forecasts, not as a matter of principle, but presumably through a lack of finances. Here forecasting was still a field of interest for individual scholars with a variety of scientific backgrounds.

In the 1930s the core of the demographic forecasting approach was clear and was set out in statistical textbooks and manuals on the population problem. At the end of the decade even Van Zanten, retired from the Statistical Office of Amsterdam, but still a lecturer in statistics at the Municipal University, almost reluctantly decided to discuss the subject of 'the future course and structure of the population' in the 1938 edition of his manual of the statistical method (Van Zanten, 1938).[58] He still held the opinion that *'the construction of a system that helps to make a judgement of the expected future course of the population'* did not belong to the domain of statistics. In his view the Net Reproduction Rate was an adequate instrument for achieving general insight into the future course of a population in the absence of international migration. Nevertheless he felt obliged to discuss demographic forecasting, because of the extensive amount of literature on this matter and many statisticians occupied themselves with population forecasting (Van Zanten, 1938, p. 154).

It was the first introduction in a statistics manual for university students to the method of demographic forecasting in the Netherlands.[59] Van Zanten had to face the problem that, for a balanced treatment of the subject of population forecasting, urban population forecasting could not be left out. He considered the efforts of his colleagues in the Department of Town Planning of Amsterdam interesting and important from a theoretical, scientific point of view, because of the way migration was included in the calculations. For the same reason, because migration only added to its speculative character, he was less enthusiastic from the point of view of forecasting practice (Van Zanten, 1938; De Gans, 1994a; also Chapters 2 and 7 of this book). His concise text on population forecasting gives an excellent overview of the national and international state-if-the-art literature. The Tokyo papers of Gini, Jensen, Ptoukha and Thompson form the main body of the foreign literature admitted in his reading list. Other items were Kuczinski's *The measurement of population*

---

58 Burgdörfer (1935) in Germany, and Carr Saunders (1936) in England had preceded Van Zanten.

59 An explanation, for students of economics, of the principles of demographic forecasting and of methods to simplify the calculations without neglecting the main principles was given by De Jong in his introduction to the population problem eight years later. It is striking that (international) migration was not only neglected, it was completely absent from De Jong's explication (De Jong, 1946, p. 168-169).

growth (1935) and a discussion in the German Journal *Allgemeines Statistisches Archiv* (1935/1936) on forecasting methodology between Burgdörfer, one of the most influential German demographers of the 1930s and G. Paulinus, an insurance mathematician who had recently completed his dissertation on population forecasting and was in favour of short-term forecasting (Van Zanten, 1938, p. 154; Paulinus, 1934).

In comparison to Van Zanten's overview of the state-of-the-art of population forecasting methodology and literature the first manual on the population problem published after World War II by the Dutch economist De Jong (1946) is disappointing. His book set the trend of the after-war disregard of the Dutch endeavours of the preceding period. De Jong referred to three international forecasts he considered exemplary: The detailed forecasts of the *Statistisches Reichsamt* of 1930, the scrupulous forecast of Sauvy of France of 1932 and those of Charles of the population of England and Wales (1935). Of the Dutch forecasts only that of Methorst (1937) is mentioned, presumably because De Jong was mainly interested in assumption making, not in the forecasting method as such. It is astonishing that not even the innovative forecasts by Van Lohuizen and Delfgaauw of Amsterdam (1932) and Angenot (1934) of the population of the Netherlands and the Rotterdam Harbour Area were taken into account. These forecasts belong to the most interesting forecasts of the 1930s, among others because of the way migration was dealt with. About this and other innovative features of these forecasts more in Chapter 7.

Angenot's forecast is particularly interesting because of his attempt to introduce a formal, matrix approach of demographic forecasting, among others with respect to future fertility. He developed a simple model wherein observed cohort fertility rates are made dynamic, which he applied both to the national forecast and the Rotterdam variant. The rates (of ten year cohort groups) were calculated by relating the number of births of the cohorts 1901-1910, 1891-1900 and 1881-1890 in the interval 1921-1930 to the number of women of these cohorts on 31 December 1931. Next the rates were corrected for the birth rate level of the period 1927-1931. These rates served as the starting rates. The rates could be made dynamic in the forecasting period by attributing to each ten year forecasting interval a fertility coefficient k_i (i = interval i). A forecasting interval specific fertility coefficient, k_i, related the level of fertility of that interval to that of the starting interval s ($k_s = 1$). The fertility rates could be obtained by multiplying the fertility coefficient of an interval by the respective starting rate (Angenot, 1934, pp. 110-122). It was assumed that future fertility change would affect all cohorts of the fertile age groups in a similar way.

The international list of demographic forecasts was growing rapidly[60]. Featuring in such lists does not imply, however, that a forecasts is outstanding or innovative from a methodological point of view. Because of the amount of calculation work, forecasts listed were often simplified and in many cases assumption making was poor. What contemporaries found interesting was not so much the innovation of methodology but whether a new forecast tended to confirm the main trend of Western population dynamics. Also whether the international demographic tendencies and its consequences, which were pictured so vividly by Kuczynski, applied to the country for which a specific forecast had been made. The new cohort component method was also used to calculate the level of fertility needed to prevent a future decrease of population (e.g. Vulhopp, 1932).

4.11 | Demographic Forecasting and Society in the Shadow of World War II. Contemporary Reflections

As a scientific field grows towards maturity self-reflection can be expected, as exemplified intectium 4.7 by the debate on the length of the horizon of forecasts, or the speculative nature of forecasting and its use for policy purposes at the Tokyo conference of ISI. Population forecasting reached maturity in the second part of the third decade of the present century, a period rendered ominous by the shadow of Fascism and Nazism. There are not many signs that population forecasting experts were inclined to reflect on this intellectual activity and its social and political implications, for instance on

60 Carr-Saunders' (1936, figure 26) overview of "*some of the more interesting results*" of the work of statisticians in many countries contains forecasts of Great Britain and England and Wales (by Leybourne, 1934 and Charles, 1935); of Scotland (by Charles, 1936); of Italy (by Gini and Finetti, 1931); of Denmark (that of Jensen made for the Tokyo conference); of France (by Sauvy, 1932); of Belgium (by Baudhuin, 1931); of Germany (by Kahn and Burgdörfer, 1930) and of the United States (by Thompsom and Whelpton, 1933). The *Selected List of Articles and Books Containing Estimates of Future Populations or Discussions of Such Estimates* (Glass, 1940) is already too long to allow for such an individual listing. Listed are forecasts of a variety of countries, including Belgium (population forecasts of Baudhuin, 1931 and 1932; and Vulhopp, 1932) and Japan (Uyeda, 1933).

lessons learned from the strategic uses of population forecasts in the preparations for the First World War (see Chapter 3). The examples discussed below reflect the various societal and political contexts of population forecasting.

The most explicit and outspoken reflections came from Van Zanten. As the director of the Statistical Office of Amsterdam, he had frequent contacts with his German colleagues and he was invited to their yearly meetings. It may be assumed that he was familiar with the state of affairs in Germany. He was Jewish and perished in concentration camp Bergen Belsen in 1944, so one is inclined to interpret his considerations with that in mind. One must be cautious and realise that the core of his considerations in the 1930s were not very different from that in the 1920s. A Dutch Jew would find it hard to imagine that Nazi ideology was anything more than temporary insanity.[61]

In contrast with neighbouring countries the Netherlands had little active experience of war, apart from some colonial skirmishes. The last Dutch military expedition in Europe against the Belgians dated back a century. The Dutch had escaped the turmoil of the Great War, and were unprepared for the Second World War.[62] Dutch population forecasters had no mistrust or reserve towards population forecasts nurtured by the strategic demographic uses to which France had fallen victim, nor by the advocacy of the host of their German and Italian counterparts that forecasts and politics should be closely linked. Dutch population forecasters were convinced of only one thing; the speculative nature of population forecasting. As has been said, up to the early 1940s, with the Netherlands under German occupation, national population forecasting was an activity exercised by individuals interested in the population problem and the future growth of the population as a consequence of the presumed future course of the birth rate. Never in the inter-war period did forecasting become the task of governmental statistical offices.[63] Strategic

---

61 Private communication to the author from his grandson, Mr. J.H. Van Zanten.

62 *'In 1940 we were the most obtuse nation of Europe. We had no notion of evil'* (professor Willem Heinemeijer, in Folia, weekly for the University of Amsterdam, 1995, 34, p. 5).

63 P. 6 of the NCBS Annual Report 1937 refers to a cyclostyled communication on *'demography in the light of new data, among others the problem of the future size of the population'* (Arch. NCBS). The communication does not, however, contain the outcomes of a true population forecast. Calculations of the Net Reproduction Rate over the preceding years merely prompted NCBS to give a general indication of the future tendencies of the birth and death rates under the influence of the expected changes of the age structure of the population of the Netherlands (CBS, 1937). The newly calculated

demographic arguments were rarely used by the main forecasters themselves. In the Netherlands the decrease of the birth rate raised anxieties of race suicide among some scientists, but the fear rarely took on serious proportions.

Van Zanten was unrelenting about the attitude of statisticians to population forecasting: The reliability and objectivity of public statistics should be beyond doubt and should never be undermined by any active involvement in speculative endeavours of calculations of the future size of population. On the other hand, statisticians had to be open to the statistical demands of society, forecasters included (Van Zanten, 1938, p. 154). Van Zanten distinguished between scientific forecasting and forecasting for policy making purposes, town planning in particular. While exploratory in character, scientific forecasting may be sophisticated, for in that case its speculative nature causes no harm. It is a different matter for forecasts made for policy purposes, such as town planning. Here the aim is to provide information about the most likely future course of population. Such calculations are highly speculative and can mislead authorities building policy on such calculations. For that reason these speculations should be simple, or should not be made at all (De Gans, 1994a). The position taken by Van Zanten was clear but, considering the forecasting activities of his colleague statisticians, he was overtaken by events.

Van Zanten's criticisms of practical forecasting resemble those of Bouthoul, a French contemporary although, in contrast with Van Zanten, he took an anti-quantitative position. As a university professor his situation was different; he could more readily position himself at a distance since he was not a forecaster. The experience of the strategic demographic past of France in relation to Germany echoes through his views, and there seems to be a premonition of the war to come.

Bouthoul (1935, pp. 228-234) started by critisising the inclination of demographers to draw from forecasts conclusions that were much too firm. In doing this they were displaying a confidence that the future would be a continuation of the past and the present with similar characteristics. If demographic forecasting were only an innocent game of scientists it would not have to be taken seriously. Science owes its advancement to testing assumptions ('hypothèses') and in itself it carries no danger. Unfortunately, the fashion of

---

NRRs presumably induced Methorst to write his article on the pivotal significance of the age pyramid in the study of future population (see Methorst, 1937).

population forecasting coincided with a heroic period of nationalism. Bouthoul thought the offensive use of statistics to be a recent phenomenon, one of the manifestations of what he termed the 'quantitative civilisation': The tendency to prefer quantitative to qualitative values. Undoubtedly the observation of differences of quantity is the easier of the two; counting is easier than analysis. While the counting of material things has advanced greatly, the statistics of social phenomena that are less easy to observe, moral statistics in particular, had hardly begun. Nationalistic competitions tend, according to Bouthoul, to use statistics as a weapon. The great triumph of nationalism was that, in Bouthoul's view, it succeeded to incorporate even statistics, in the form of statistical predictions of the future, in its aims. Or, better still, for dictating the measures that have to be taken in advance, once one is forewarned by forecasts. If it is believed that through statistical forecasting it is known what the population will be within, say, fifty years, then wisdom commands to act accordingly in order to prepare for the future situation that is taken for granted (Bouthoul, 1935). In effect Bouthoul's critical evaluation of population forecasting is a reflection of the traditional negative French attitude towards forecasting, dating from the end of the 19th century. On the other hand, it is a plea to consider forecasting as an activity taking place within a societal context to be taken seriously as such, for instance by an investigation of its strategic demographic dimensions.

How different was the situation in Italy and Germany. From the forecast of 1926 onwards, demographic forecasting became the task of the German National Statistical Office in Berlin. Like Gini and other leading Italian students of population many outstanding German statisticians propagated a close relationship between population forecasts and population policy.[64]

German statisticians such as Burgdörfer and Benser did not show any of the reserve towards the 'speculations' of demographic forecasting propagated by

---

64 It is beyond the scope of this investigation to explore the relationships between population and policy in general and Fascist and Nazi policy and population and its consequences in particular. For an investigation of the relationship between fascism and population in a comparative European perspective, see Weindling (1989); of the part played by the medical profession in Germany, Weindling (1990). For an investigation of abortion and eugenics in Nazi Germany, see David *et al.*, (1990). For a concise study of official population forecasting in Gèrmany since 1925 and its relationship with policy, see Höhn (1986). With respect to Italy see Ipsen's book *Dictating Demography. The problem of population in Fascist Italy* (1996).

Van Zanten at Tokyo. To Burgdörfer the old method limiting itself to the establishment of what 'has been' and of what 'is' was no longer sufficient; one should turn to a biological point of view to establish a continuing basis for what 'will be' (Burgdörfer, in: Paulinus, 1935/1936, p. 150).

Most of the time statisticians abstain from general opinions and give a neutral and objective description of method and assumptions. That does not mean that their writings were free from 'coloured' or 'strategic demographic' points of view. The forecasts of the national population of Germany from 1926 onwards were not (Höhn, 1986). Already the first demographic forecast, based on the assumption of either a constant number of births over the next 55 years (at its best) or a continuing decrease in the general fertility rate, had brought the possibility of a decrease of the size of total population in the future into the focus of public opinion. The Nazi authorities of later years had no problem in interpreting the 1928 forecast, which bore the sober title *'View of the future population development of the German "Reich"'*, as a political call. Demographic forecasts became in retrospect political instruments, even means of propaganda (Höhn, 1986, p. 215). Presumably, this was in line with the opinion of leading statisticians. Benser for instance, most expressively stated that population statisticians should use the method of projective calculations ("*Methode der fortschreibenden Vorausberechnung*") both in practical work and in demographic research. Mathematics were of little help in the solution of the population problem. Forecasts should be closely linked to the vital processes. For that reason, forecasting methods should take into account the three main vital processes, birth, reproduction and mortality, that correspond as closely as possible with the real life course of generations (Benser, 1940, p. 282). This method only gives insight into the biological condition of the population ("*die volksbiologische Lage*"), in the direction of population policy to be taken, and at the targets of such policy. The direction of the future course of population depends on the objectives of the forecast and can originate from demographic, economical or military considerations.

An indefatigable inter-war protagonist of the fear that Germany was to become a nation without youth ("*Volk ohne Jugend*") was Burgdörfer (1935), director of the German national statistical office. Given the low level of the net reproduction rate in comparison with other countries, his fear is understandable. It is more difficult to comprehend how he was able to combine this fear with the belief that Germany was a nation without sufficient physical space ("*Volk ohne Raum*").[65]

But even in Germany divergent positions were taken, for instance by a young actuary, Gottfried Paulinus, who dared to defy Burgdörfer (Paulinus, 1934; 1935/1936). Paulinus was a critic of Burgdörfer's analytical-demographic assumptions. He blamed his inclination *'to cite his figures as if resting on foundations that are absolutely accurate'* and questioned the feasibility of a pronatalistic policy by the state, starting from the assumption that the decision to have a child belongs to the domain of the will of the individual.[66]

Assumption making in long term population forecasting, based on expectations with respect to the future of population, is closely interrelated with population theory and with beliefs about the future of society. In many instances it was scientific interest in the future course of population resulting from a continuation of observed demographic trends that induced inter-war economists and statisticians to make forecasts of the national population. This is particularly the case for Dutch forecasters. The idea of using national population forecasts for government policy purposes was absent. Moreover, the future course of population was considered to be a datum (Ter Heide, 1992; De Gans, 1995b). The idea of actively influencing the future course of population, particularly mortality and fertility, was almost absent. It was different with German

---

65 Burgdörfer was the originator of the concept of the "*Bereinigten Lebensbilanz*" (the 'adjusted balance of life') in 1929, an alternative to Kuczynski's NRR. It was founded on stable population theory. It enabled Burgdörfer to determine the "*Geburtensoll*" (the number of births that **should** be achieved), defined as the minimum number of births needed for the permanent safeguarding of the national population total; also as the number of births that safeguard the continuance of the active part of the population (the "*produktiven Volkskraft*") (Benser, 1940, p. 280). Methorst's population forecast of 1937 was founded on the concept.

66 "*Immer wieder zitiert Burgdörfer seine Zahlen, als ob sie auf den sichersten Grundlagen beruhen.*" (Paulinus, 1934, p. 133). Paulinus was the son of clergyman, and had studied insurance sciences, statistics and physics at the universities of Freibuth and Leipzig. At the age of 27 he had finished his thesis on population forecasting. He built partly on criticisms of Würzburger, who had questioned Burgdörfer's explications of the recent fall in the number of births in Germany, and Von Mises (1932), who had pointed out that Burgdörfer's predictions, insofar as they rested on stationary population theory, would take an infinitely long time to become reality. Paulinus agreed with 'the majority of statisticians, including Von Mises and the Swedish statistician Sven D. Wicksell, that population forecasts with a horizon of more than one or two decades were useless (Paulinus, 1934).

forecasts. They were intended to incite state intervention that would result in a change in the future course of fertility. The coercive power of forecasts in combination with the course of events in Germany worried Bouthoul and brought him to advocate the renunciation of forecasting. The speculative nature of long term population forecasts led Paulinus and Van Zanten to advocate that only short term forecasts should be made and –in the case of Van Zanten– to suggest that, if making forecasts for policy purposes could not be avoided, they should be as simple as possible; in case of a future change of tendency of the observed trends a new forecast could easily be made (De Gans, 1994a). Van Zanten's position with respect to population forecasting by statisticians in the years that lay under the shadow of the Second World War, namely that statisticians should completely refrain from forecasting because it could endanger the credibility of their work, can also be seen as a token and an example for the contemporary and future forecasting profession to safeguard the objectivity and independence of its population forecasts.

4.12 | Clash of Methods or Clash of Paradigms?

The period between the two world wars can be divided into two separate parts. The part, which coincides with the 1920s, is the period of the struggle for paradigm dominance, because of the (re)emergence of demographic forecasting and the (re)kindling of the belief in the existence of a law of population, due to the (re)invention of logistic growth methodology. The emergence of demographic forecasting took place almost simultaneously in several countries. While in the Netherlands demographic forecasting resulted from a national debate on the population issue of a population that was growing too fast, in other countries, England in particular, demographic forecasting resulted from fears of a continuation of the fertility decrease.

In the 1920s, demographic forecasting had to prove it had come to stay. The mathematical-biological concept of logistic population growth had the hallmark of true science, not only because of its expression in formal terms but also because it originated in biological science. The debate on the future of population at the Royal Statistical Society in 1924 and at the first World Population Conference of Geneva 1927 noted the merits of logistic growth with all its law-like attributes in comparison with those of demographic forecasts. Mathematical 'law' appeared to supersede 'speculation'. By the beginning of the 1930s the battle had been won by demographic forecasting. The Tokyo conference of ISI in 1930 and the Rome conference on population studies (1931) marked demographic forecasting. In Tokyo, statisticians discussed the issue of responsibility: Should statisticians be involved in population forecasting; should forecasts be used for policy purposes and should forecasts have long-term or short-term forecasting horizons. The Rome conference resulted in recommendations to governments to use demographic forecasts.

The second part, coinciding with the 1930s, is the period in which demographic forecasting became the new standard for the calculation of the future development of national populations. The number of forecasting experts was growing fast. Methodological issues of forecasting were discussed at national and international meetings of statisticians and students of population. In some countries, though not yet in the Netherlands, forecasting became the task of public organisations, statistical offices in particular. In various countries, the Netherlands included, demographic forecasting methodology was treated in textbooks for university students.

The top of the statistical establishment of the Netherlands, represented by the statisticians Methorst, Verrijn Stuart and Van Zanten, stood in the heat of the international battle on forecasting methodology; they, not the Dutch pioneers of demographic forecasting methodology, had positions to defend and reputations to lose. These men were, or had been, heads of statistical offices. They were delegates to international conferences where the issue of future population and its calculation was at stake. They were not afraid of intervening at international meetings, or contributing to international journals. Although not all of them were forecasters, they had an overview of the national state of the art of population forecasting. The advancement of national interests in the international field was, presumably, never out of their mind.

A clash of methods or a clash of paradigms?

The most striking feature of the debate of the 1920s is the absence of attempts to arrive at a rational weighing of the pros and cons of the methods under discussion. Apparently, the participants in the debates could only opt for the position of either a supporter of or an opponent to one of the methods involved. There was no way in between. It demonstrates that the struggle for dominance was much more than a mere struggle between methods. The controversies went much deeper and touched upon fundamental beliefs of reality. For that reason one is closer to the mark if it is assumed that basically a clash of paradigms was going on in the world of statisticians and statistics oriented students of population.

The assumption accounts also for the behaviour of the leading 'internationals' among Dutch statisticians. Verrijn Stuart and Van Zanten felt strongly that 'speculative' (demographic) forecasting was not part of the domain of statistics. For this new field of quantitative scientific activity was omitted from the new editions of their statistics manuals in the 1920s (e.g. Verrijn Stuart, 1924, 1928; Van Zanten, 1927). It was not in their interest to take an international stand with respect of the achievements in population forecasting of their compatriots. Verrijn Stuart nor Van Zanten could 'fit' demographic forecasting in their conception of statistics. It excluded demographic forecasting as an activity to be taken serious in the context of debates among the members of their profession. Van Zanten in particular stood for his principle that statisticians and statistical organisations should completely refrain from forecasting, even to the extent of preventing a true debate on forecasting methodology in Tokyo.

The debate brought Methorst, who had a positive attitude towards demographic forecasting, in an uncomfortable position. Methorst lacked the mathematical background to permit him to be actively engaged in debate between the international giants of mathematical statistics and mathematical biology. He manoeuvred to make his point, but outside the central arena of the debate and without taking publicly sides. In such a way of operating there was no place for the advancement of knowledge about the achievements of his countrymen. Moreover, Methorst was always in the habit of referring to the contributions of his peers in the international statistical society, not to the performances of men who took only marginal positions in statistics, as Wiebols and Oly, or complete outsiders, as 't Hooft.

The international recognition of the innovative achievements of Dutch population forecasters fell victim to the positioning of the 'internationals'. The pioneers of the modernisation of population forecasting in the Netherlands could not contribute to the international propagation of the new methodology. They lacked the proper academic and professional positions and they were not in the position of being delegated to international conferences where the methodological aspects of forecasting were at stake. In the next chapter it will become clear that they contributed actively to the national debate on the proper method of calculating the future of population, which was their main focus. Making contributions to international scientific journals of their achievements was beyond their scope.

5. Competing Methodologies in the Netherlands

5.1 | Standard Geometrical Forecasting in the Point of Departure

At the end of the First World War, geometrical extrapolation of total population was still the usual approach; that was the case both in the Netherlands and elsewhere. The demographic predictions of Cannan, Fahlbeck and Westergaard produced in the preceding period were isolated occurrences. These forecasts had not yet found a direct following and it may well be assumed that knowledge of their innovations was restricted to a small circle of economists and statisticians. Methodological innovation had to depart from geometrical extrapolation. The 1920s and 1930s were to develop into the period in which modern population forecasting reached its full growth. That demographic forecasting was to become the new standard forecasting method of the national population could not have been predicted from its pre-war history. But the promise of its development was already present in the few well founded examples of geometric population growth forecasts: Rückert's municipal population forecast published in 1917 (which is discussed in Chapter 6) and Rooy's national population forecast of 1921. Rooy's forecast is of particular interest because it had to do with a societal core issue of the post-war period: The housing shortage.

The issue of the housing shortage was seen as a serious problem and raised public anxiety. Opinions differed whether the shortage could be solved in the near future, say within ten years, or had to be seen as a structural problem. The policy of the Minister of Labour, responsible for housing, was founded on the conviction that the housing shortage problem would soon be solved (Rooy, 1921). Critics of governmental expectations, such as Faber (1918), Bloemers (1920) and Rooy (1920; 1921) had quite different views. According to Bloemers (1920, p. 45), many believed that, once mobilisation was over and the repatriation of (mainly Belgian) fugitives completed, the housing shortage would disappear. These people were surprised when they discovered that the housing need was more acute than ever during the war. Part of the housing shortage at the end of the war resulted from the increase of population. The direct effect of the war on this part of the housing shortage was limited; it merely aggravated the housing need dating from the pre-war period. The war only made it clear that the national housing stock was very restricted.

It was Rooy who provided the critics of government housing policy with quantitative arguments based on an analysis and forecast of the determining factors of the housing shortage. He warned against the dangers of underestimating future housing demand and blamed the government for not taking sufficient account of such factors as the negative effects of the eight-hour working day on housing production, the effects of the increase of population, and the decrease of average household size on future housing demand. He concluded that an emergency situation would arise in the near future; an emergency only few people were aware of.

Future housing need was calculated in a simple, straightforward and pragmatic way, with extrapolated average dwelling occupation figures. Changes of population composition are not accounted for, but the approach is still valid.[57] Rooy's population forecast was based on geometrical population growth and a constant ten year growth rate of 16.6 per cent. The method is simple but satisfactory, given the task Rooy had set himself. It is an example of everyday forecasting practice. The plausibility of the assumption of a constant future growth rate is sustained by demographic arguments, deduced from observed tendencies in the birth and death rates keeping their balance. In everyday forecasting practice the Malthusian notion of a law of population growth clearly belongs to the distant past.

[57] The arguments of the critics of government housing policy are strikingly similar to those of the 1950s and 1960s. The debate arouses a *déja vu* sensation in this respect.

Vignette 9: Rooy's calculation of future housing need of 1920

Rooy started by estimating the housing shortage in 1921. On the basis of the decrease in the average occupation figures from 4.54 persons per dwelling in the 1899 census to 4.47 persons per dwelling in 1909 and an assumed linear continuation of the decrease from 1909 to 1920, he estimated that the figure would fall to 4.4 in 1920. The population total of 1920 was estimated at 6.7 million. The housing need could be estimated at more than 1.5 million dwellings, which would have meant an increase of 255,745 houses in the 1909 housing stock. This figure did not include the replacement of obsolete dwellings. To satisfy the calculated increase in housing demand, an average total of 23,250 houses per year should have been built in the period 1909-1920. However, the number of inhabited dwellings had only increased by an average of 17,800 per year. The absolute housing shortage in 1921 could be estimated at about 125,000. Taking into account the working off of arrears in the necessary replacement of obsolete houses from 1914 and the necessary stock of uninhabited houses (estimated at three per cent of the total housing stock), the period 1909-1920 had seen a deficit accrue of 165,000 new dwellings, seven times the current average yearly housing production. Mainly because of the First World War, housing production equivalent to the total for seven years had been lost.

Once the housing shortage in the year of departure of his forecast had been assessed, Rooy could proceed to a forecast of the increase in housing demand resulting from future population growth. Assuming that future decrease in natality would be more than compensated by future decrease in mortality, he expected a continuation of total population growth in the next two decades at the level of the ten year growth rate of 16.6 per cent in the past decade. The population was estimated to reach 8.08 million inhabitants in 1930 and 9.42 million in 1940.

Rooy assumed a constant proportion of persons not living in houses (3.3 per cent of the population total), and a continuing fall in the average dwelling occupation figure by 0.07 percentage points per decade. He calculated that, between 1921 and 1941, another 615,000 houses would have to be built, resulting in a total of 740,000 new houses (125,000 + 615,-000). The stock of uninhabited houses had to be set at three per cent of the total housing stock (65,000 houses). Obsolete houses had to be replaced, as had dwellings lost by city formation. Taking all these factors into account, actual future demand could be estimated at another 15,000 houses per year (300,000 in twenty years). In total, 1,105,000 dwellings would have to be built in the next twenty years, an average of about 55,000 per year.

With the means currently available, a maximum of approximately 25,000 houses per year could be built. Rooy came to the disconcerting conclusion that the necessary yearly production fell short by 30,000 dwellings (De Gans, 1997).

5.2 | The Context of Innovation: Rapid Population Growth

The incentives for innovation came not from the needs of planning and decision making, but from substantial social and scientific interest in the population issue. The interest was stable throughout the inter war period, even though it had many facets and took many forms. The interest shown in the problem and its consequences for national welfare was the catalyst for interest in the future course of the population of the Netherlands and therefore also for the way in which future population could best be calculated.

In contrast with other West European countries, the population issue appeared quite late on the social and scientific agenda in the Netherlands. This is not to say that interest in the population issue did not previously exist. It did, as is clear from the reactions to the demographic situation verbalised by people like Pierson, Verrijn Stuart, Methorst and Saltet and Falkenburg in the preceding chapters.

Nevertheless, only the beginning of the 1920s seemed to be ripe for a national debate. The stagnation of the Dutch economy following the short spurt of growth directly after the end of the First World War, the subsequent unemployment and the accompanying poverty, caused economists to fear that the law of diminishing returns would set limits on progress. '*No century could experience a relative increase in social income of a size, expressed in consumer goods, which could be equal to that which was achieved in the last one hundred year*' prophesied Verrijn Stuart (1919, p. 81). It was imperative in his view that the continuing growth of the Dutch population should be stopped. The topic gained attention as soon as it was introduced and remained high on the scientific and societal agenda for the next few years.

Those who shared Verrijn Stuart's views had the Neo-Malthusian Society (Neo-Malthusiaanse Bond NMB), founded in 1881, on their side. Where marriage and the family were concerned, the NMB aspired to achieve a satisfactory balance between the means of existence and the size of the family in order to reduce poverty (Van Praag, 1977). The methods the Neo- Malthusian Society propagated, such as intervention in fertility through artificial means so that the number of births could be controlled, were considered by many to be evil. As far as the opponents of the ideology were concerned, the real core of the population issue lay in the propaganda of Neo- Malthusianism itself and not in a population surplus. The difference of opinion concerning the issue at stake did not help make discussions on 'the population issue' clear (Van Praag, 1977).

The NMB met with fierce opposition from its inception. Its opponents came principally from the circles of the confessional political parties of the day and the religious groups supporting them. The leaders of the social democrats were more inclined to give cautious support to pronatalist policies, or to take a neutral position, rather than support antinatalist policies; those in socialist circles who did not already disapprove of the NMB had in any case distinct reservations. The supporters of the NMB could generally be found in liberal and liberal democratic circles. As much individual freedom and limitation of state influence as possible were of paramount importance to them (Van Praag, 1977). In the first few years of its existence, the NMB's propaganda was virtually limited to intellectuals. It was not until 1900 that the Neo- Malthusians started to concentrate on the mass population. The discussion of the issue of overpopulation began late in comparison with other countries like England (Soloway, 1990). This had to do with the similarly relatively late commencement of industrialisation and urbanisation and the relatively large influence of religion on public opinion in the Netherlands (Van Praag, 1977).

Later, in the 1930s, the population issue again became important, but this time in a totally different form. Attention was focused not on the increasing population, but on the declining birth rate. The continuing decrease in this rate led to social Darwinian concern for the survival of the population, but also for the quality of the population to be maintained.

5.3 | The 'Great Neo-Malthusian Debate' and its Quantitative Foundation

It will now be clear that, until the beginning of the 1920s, little had changed in terms of practical population forecasting in the way in which the size of the future population was calculated, particularly not in the Netherlands. The conviction that there was a law of population to which a population was subject was no part of assumption making in everyday forecasting practice, as is demonstrated in the case of Rooy's forecast. Neither did this mean that Malthusian law was rigidly adhered to; it is hard to find examples of such a belief in the Netherlands in the first part of this century.[58] In general, the approach of geometrical population growth is pragmatic. It is just a method to be used and in many instances it is accompanied by an analysis of observed trends of the components of natural population growth.

Since censuses were held approximately every ten years from 1829 onwards, the population growth between two successive counts could be calculated. It was therefore known that the average annual growth rate in consecutive inter-census periods was not constant, but was subject to change. Paradoxically, it was a calculation of the geometric type in its most orthodox form that made a great impression in the early 1920s, even to the extent that it became an indirect cause of the methodological shift that ended the dominant position of geometrical forecasting. Its impact came not so much through the quality of the methodological basis, but rather through good timing and the authority of its maker, Verrijn Stuart (Verrijn Stuart, 1919; 1921). As secretary and treasurer of the Society of Political Science and Statistics (Vereniging voor Staathuishoudkunde en Statistiek VSS), Verrijn Stuart was its driving force. Thanks to his efforts, VSS stirred up a national debate on the population problem. At the annual general meeting of VSS in 1922, many of the direct and indirect contributors to the innovation of population forecasting in the Netherlands were brought together for the first time.

Verrijn Stuart was a student of the statesman and liberal economist N.G. Pierson. As an editor, Verrijn Stuart had revised the 1913 edition of Pierson's manual of political economy (Pierson, 1890/1913). In the wake of Thomas Malthus, Pierson was convinced that insufficient production and high natality should be recognised as the main causes of pauperism. In his eyes the national birth rate was too high; no real improvement of the prevailing economic conditions could be expected while natality failed to decrease. Immoderate natality would result in poverty and distress and would end in immoderate mortality. Like Malthus, Pierson opposed rash and frivolous marriages among the poor; marriages based on passion rather than good prospects. Rather sarcastically, he ridiculed the views of those who thought poor people, married, with many children and living in abominable circumstances, as thereby possessing poetic beauty (Pierson, 1876, p. 24). Although Pierson had

---

58 The conviction did not disappear altogether. It was kindled again with the introduction of logistic growth methodology. Outspoken adherents in the Netherlands were, for instance, Lewandowski and Van Dranen (1933).

difficulty in siding with Neo-Malthusianism with respect to the advocated practices of family limitation, he also had difficulty in denouncing it (Pierson, 1976; 1890/1913). Verrijn Stuart shared Pierson's opinion that the Dutch population was growing too fast and was a restraining factor on the growth of national income; the law of diminishing returns placed a limit on possible progress (Verrijn Stuart, 1919).[59] He was more outspoken than Pierson that population growth should be halted with all means, including those propagated by the Neo-Malthusians.

The calculations of Verrijn Stuart amounted to the following: If the future growth rate were maintained at the average level of the period 1909-1920, at 1.42 per cent per year, the population would double in 49.16 years. In one hundred years' time the population of the Netherlands would be 27.5 million and in two hundred years 110 million (Verrijn Stuart, 1921; also 1919; compare Pierson, 1876; 1890/1913).

Besides Pierson, Verrijn Stuart succeeded in putting forward for public discussion a number of economic problems connected with a fast growing population (Van Praag, 1977). He was convinced that, if the birth rate maintained the high level of the last few years, an increased death rate and 'proletarisation' of the population would be inevitable. Verrijn Stuart was an advocate of curbing fertility.[60] By taking this position, he opened up in the Netherlands what was later called 'the great Neo-Malthusian debate' (Van Praag, 1977).

The results of Verrijn Stuart's calculations became lively topics in the debate on the population issue (the causes, consequences and combating of sharp population growth) that was the central theme for the annual general meeting of VSS of 1922 and –as a reaction to this– of the Roman Catholic Dutch-Flemish Congress on the population problem held in Breda in 1923.

---

59 Van Praag (1977, p. 253) is not correct in assuming that Verrijn Stuart was the first to discuss economic problems associated with a fast growing population. Pierson was ahead of him.

60 In 1926 he put pressure on the editorial secretary of *De Economist* to ensure that, in a text supplied by 't Hooft, the emphasis would be placed on the importance of a reduction in the birth rate for combating overpopulation and therefore eliminating its terrible consequences (letter to G.M. Verrijn Stuart, Arch. of *De Economist*, file nr.1926/28-'t Hooft).

The meeting of the VSS was orchestrated by Verrijn Stuart in his capacity as secretary. He may have been instrumental in setting the theme. It was Verrijn Stuart who invited the authors of the preliminary reports to the meeting. It was he who, even he did not dictate the subject, content or composition of their contributions, at least made suggestions (Kohlbrugge, 1922). Kohlbrugge, an ethnologist/physician professor at the University of Utrecht, and Ries, a jurist who had recently contributed an article on Malthus to *De Economist* were invited to prepare preliminary reports. They were asked to discuss the issue of the impact of the economic developments of the last few years on the population figure, particularly in the Netherlands. Both Kohlbrugge and Ries were influenced by social Darwinism, but in different ways.

Kohlbrugge saw the population issue as a short term problem. He attributed unemployment and hunger in Europe to the war and the chaos it had produced. He saw no reason to become alarmed about overpopulation. He therefore did not consider it necessary to combat population growth. However, he was convinced that Neo-Malthusianism should be combated, considering it evil, which, should it proliferate, could result in the depopulation of the Netherlands. In his preliminary report, Kohlbrugge took such an oversimplified position that both during and after the meeting, he received quite a few critical comments, from Verrijn Stuart included. Kohlbrugge later wrote that his stance against Neo-Malthusianism stemmed from a misunderstanding, as he had understood that this was why he had been invited by Verrijn Stuart (Kohlbrugge, 1922).

Kohlbrugge's defence must be taken with a grain of salt. Extreme attitudes to Neo-Malthusianism appealed him. His antipathy is still apparent in his textbook on practical sociology written in 1929. In this book he expressed the opinion that unrestricted Neo-Malthusianist influences would have adverse consequences of a physical and ethical nature for the population and would result in economic debilitation of the population. If degeneration could be eliminated from the population, then the future of Europe would belong to us. Should our children and grandchildren not be able to eke out an existence, then he would prefer famine, epidemics and war to the unrestrained propaganda of Neo-Malthusianism that was causing the Dutch people to degenerate (Kohlbrugge, 1929).

The other preliminary advisor was L.A. Ries (1893-1962), a former student of Verrijn Stuart who had followed his courses on economy in the University of Groningen, and was a friend of his son. Verrijn Stuart had a high opinion of Ries dating from the period of his study of law in Groningen (Henssen, 1994, p. 215).[61] In the year preceding the VSS annual meeting Ries had contributed a study on Thomas Malthus to *De Economist*, where Verrijn Stuart was an editor (Ries, 1921).

Ries, like Kohlbrugge, was influenced by social Darwinism. As far as Neo-Malthusianism is concerned he can be placed in line with Pierson and Verrijn Stuart. Ries dismissed Malthus' law as a valuable part of economics as far as it rests on the assumption of the unlimited capacity of the human race to procreate (VSS, 1922, p. 99). From an analysis of Malthus' writings, Ries concluded that the problem of Malthus was not overpopulation, but pressure by the population against the limits of its subsistence; there would always be a tendency towards overpopulation. The pressure could be lessened, but never lifted. Consequently there is always fear of increasing poverty and future starvation in a growing population, if not of future wars arising from population growth. Ries was convinced that the main cause of the First World War was the overpopulation of some countries (Ries, 1921).

In his lecture Ries focused on the contribution of the natural sciences, biology in particular, to population science, for he was of the opinion that most students of population doctrine, originating from legal and theological circles, tended to neglect its findings that procreation is to a high degree a phenomenon of adaptation. Limitation of the number of children is not a natural law, but a complicated cultural phenomenon; adaptation to a changing environment is the dynamic part (VSS, 1922, p. 97). Ries saw himself as an opponent rather than an advocate of Neo-Malthusianism which as such was a great evil, but, given the economic situation, a bitter necessity (VSS, 1922, pp. 101-102).

Ries' preliminary report was far more balanced than that of Kohlbrugge. Moreover, it was based on some understanding of demographic developments. Ries was apprehensive of possible underpopulation in the long-term, too, should the decline in the birth rate continue. Nevertheless, the Netherlands

---

61 Ries was a man of great intellect, with an analytic mind, not easily impressed by the high reputation of an opponent, witty, presumably without too much consideration for the feelings of others. At the time of his lecture, Ries was on the way up of a fast and brilliant career as a civil servant. He would become a comptroller general at the Treasury in the early 1930s. Soon after, in 1936, his career ended in a sudden, undeserved downfall (Henssen, 1994).

should try to learn to live with low birth rates, for there was no way back (Verslag, 1922; Van Praag, 1976).

5.4 | A Validation of the Quantitative Foundation

The publication of the preliminary reports of Ries and Kohlbrugge in preparation for the annual meeting cast its own shadows. With a good sense of timing, Methorst had published the results of his calculations as a reaction to those of Verrijn Stuart shortly before the VSS meeting. In fact, Methorst produced the first component forecast in the Netherlands, based on separate extrapolations of the components of the rate of natural population growth.He only attempted to discount changes in the age structure verbally. The question of how far Methorst was indebted to Fahlbeck is answered in Chapter 3. On the other hand, Methorst's forecast was also close to that of Rooy, with Methorst giving a graphical picture of what Rooy verbalised.

In addition to Rooy, Methorst made a long-term extrapolation of the components of natural population growth. He showed that it was improbable for the natural growth rate to remain unchanged and that a stable (stationary) situation was more likely to arise. After all, the death rate had already dropped to a minimum, while the birth rate was still showing a tendency to decrease. The long-term results would be a change in the age structure in favour of the oldest age groups, and this would lead to an increase in the crude death rate. According to him the ultimate total population reached in 2009 would not exceed 15.5 million (Methorst, 1922).

It is interesting to note that the calculations of Rooy, although quite dramatic with respect to the estimation of the increase of the housing shortage the Netherlands was going to face in the next few decades, were not taken into consideration in the debate on the population issue. The neglect can hardly be explained from a difference in methodological quality. Presumably, Rooy's forecast did not play a part because it dealt with a problem of practical government policy, perhaps because Methorst and Verrijn Stuart were well positioned men in the world of economics and statistics, but not in the world of housing and housing reform. Moreover, Verrijn Stuart's 'science fiction' style calculations were more dramatic, while those of Methorst, convincing through their graphical presentation, could be taken as an antidote to the shock effect Verrijn Stuart intended. In this play between opposite forecasting results there was no need for the results of a third forecast.

The calculations of both Verrijn Stuart and Methorst were treated as actual forecasts. It seemed not to matter that Verrijn Stuart had never meant his calculations to be so interpreted and that he only wanted to demonstrate the impossibility of such a high growth rate over such a long period of time (Verslag, 1922; Verrijn Stuart, 1922; 1925). It is surprising that even Methorst interpreted Verrijn Stuart's calculations as a true forecast. This could be because Methorst and Verrijn Stuart were opponents in the great Neo-Malthusian debate. At that period, Methorst held an anti-Neo-Malthusian point of view (Methorst, 1922; Van Praag, 1976). In the words of Verrijn Stuart: *'Mr Methorst belongs (....) to the opponents of those who –indeed not out of pleasure– again and again mean to have to point at the necessity of a further decrease of the birth rate'* (Verrijn Stuart, 1922, p. 901).

All things considered, Verrijn Stuart and Methorst had similar aims in mind. They wanted to demonstrate that in the long run population growth would have to come to an end. Seen in that light, both approaches were effective: Verrijn Stuart chose an indirect approach, while Methorst demonstrated that a decrease in growth was actually on the horizon. From a methodological point of view, Verrijn Stuart's contribution was scant. From what he knew of Fahlbeck and Westergaard he could have done better. The quantitative foundations of his message did not differ much from those of Pierson in 1876. Although he would have been the first to be capable of conveying a better understanding of population dynamic processes to his public, Verrijn Stuart let the chance slip through his fingers.

It is remarkable that neither Verrijn Stuart nor Methorst indicated the relative value of working with crude rates, whether this concerned the crude growth rate (Verrijn Stuart) or the crude rates of the components of natural population growth (Methorst). Verrijn Stuart was well informed about what took place nationally and internationally in the population statistics field and the development of demographic analytical instruments. He was familiar with Körösy's method of working with standardised rates by means of standardisation of the age structure; he knew of the 1879 efforts of Richard Böckh, Director of the Office of Statistics in the city of Berlin, who calculated rates that were less dependent on the influence of the age structure of the population, the concepts of which later became known as the total fertility rate and the net reproduction rate. Verrijn Stuart was familiar with life table methodology. He did not believe in a fixed unalterable law of mortality. Time and again, it transpires that Verrijn Stuart was not really interested in an estimate of the actual future development of the population. He never thought it necessary to make essential corrections to the passage on the increase in the population in the new 1924 edition of his manual on the foundations of public housing of 1920. The text is similar to what he had written in articles of 1919 and 1922 (Verrijn Stuart, 1919, 1920, and 1924). The position taken by Verrijn Stuart is not surprising; given his view that population forecasting had little to do with statistics. Time and again he contented himself with warning society that population growth could not continue much longer in the way it had done before (see Verrijn Stuart, 1925). The stimulation for improvement of forecasting methodology therefore had to come from other sources.

5.5 | The Outcome of the Neo-Malthusian Debate: The Necessity of a Better Understanding

At the annual meeting of the VSS it was pointed out from various sides that a better understanding of the factors influencing the population development was much needed. The comments of Van Zanten (an old-fashioned statistician of the Dutch school) and Holwerda (one of the few Dutch representatives of the school of mathematical statistics) had a particular influence on the ensuing innovation of forecasting methodology. In order to achieve a better understanding of these factors, what was needed was better statistics, especially statistics which would allow for an in-depth study of the 'birth phenomenon', as Van Zanten stated (Verslag, 1922, p. 31). Following Ries' comments on the causes of the decrease of natality, Van Zanten wondered whether the observed decrease could exclusively be attributed to a purposeful limitation of the number of births (op. cit., p. 33). In line with A.B. Wolfe's criticisms of the international literature of economists and sociologists on this subject in the first decades of this century, Van Zanten criticised Kohlbrugge for having contented himself with the use of general statistical data provided by NCBS that could only indicate that natality was decreasing. Once the conclusion was drawn, to find an explanation of the observed decrease '*one proceeds to a general form of reasoning and without having data to check this way of reasoning to the facts*' (Verslag, 1922, p. 30). In Van Zanten's view the phenomenon of natality should be investigated much more thoroughly before conclusions could be drawn. The focus is then not on only one cause.

Fundamental suggestions for improving forecasting methodology came from Holwerda. In his view the statistician should be counselled on the relevant questions to be illuminated. According to Holwerda, Methorst's calculations had already shown what was possible if statistical data were used in a correct and useful way. Holwerda did not believe it possible to make an objective study of the population issue without adequate statistics. In order to discover why the birth rate was decreasing, one should know the following: "*There are 10,000 births, how many offspring will they produce in the long run? The following questions should then be asked: How many will reach the marriageable age? This is therefore a question of mortality; how many will marry from the number that do reach a marriageable age; what is the duration of the marriage and how many children come from these marriages of variable duration. This is the problem of marital fertility which, as we know, depends upon such factors as the length of the marriage, the age at which people marry and on the difference in ages between the man and the woman. This leads to a whole series of probability systems and if we knew exactly what they all were, we would at least know what the situation of today would be*" (Holwerda, in Verslag, 1922, pp. 48-50). In Holwerda's view, one should first establish what the present day sets of probabilities look like. Only then would it be worthwhile discussing the possible future changes within. For the time being there was no choice but to study the population issue "... *with a deficiency of statistical information*" (Holwerda, in: Verslag, p. 50).

In this way Holwerda showed the direction, for the highest geographical level of scale at any rate, that would eventually lead to a definite break in the prevailing practice of making future calculations. Holwerda became the direct source of inspiration of Wiebols (see Chapter 1). The development of demographic forecasting was not following a linear line of methodological improvement. The Netherlands escaped a true debate between the advocates of logistic forecasting and demographic forecasting, but saw a combat between methods of demographic forecasting in stead.

5.6 | 't Hooft's Confusing Theory of the Conveyor Belt...

Wiebols was not the only person to gain a following in the inter-war period. Soon his forecast had to compete with that of a complete outsider in the field, the mechanical engineer F.W. 't Hooft. This man not only supplied an alternative to the Wiebols method, but over several years he left an impressive mark on the discussions on the way in which the future size of the population of the Netherlands could best be calculated. Thanks to ‘t Hooft it is possible to gain a clear picture of the analytical-demographical tool-kit of population forecasting methods at that time and its use in population forecasting. He can be seen as an important catalyst in the distribution of the existing knowledge and perceptions in this field. At the same time he caused considerable confusion by obstinately holding to some incorrect perceptions resulting from parallelist beliefs (as discussed in Chapter 3).

The nature and meaning of his work can best be understood if one is aware of the principal concept behind his work. As an engineer he was interested in social issues. As far as the population issue was concerned, he believed that better results could be achieved through calculations than he had seen until then. In 1913 his ideas took shape when he was present at the testing of a conveyor belt. His ‘conveyor belt theory’ (see *Figure 5.1*) became the foundation of his approach of population dynamics (‘t Hooft, 1929, V).

The metaphor runs as follows: Population dynamics is a process similar to that occurring on a conveyor belt. Granules (births) are thrown onto one end of the belt. Some of these granules disappear during the transport of the belt (deaths at a young age) and some of them are thrown off at the end of the conveyor belt (deaths at the end of the ‘natural’ span of life). This conveyor belt can be replaced by a conveyor belt with a length equal to that of the average life span of the deceased: Now the granules thrown onto one end of the belt will all fall off simultaneously at the other end of the conveyor belt.

The size of this substitute population is equal to the product of the number of births and the average age at the time of death, not to be confused with the average life expectancy or the average age of the population. If the conveyor belt is extended, that is to say if the average life span of the deceased increases, then the belt can accommodate more granules at the same time: The population increases, as a result of the lengthening span of life. In time, however, the balance will be restored, after which the size of the population, being again the result of the product of the (new) average life span of the deceased and the number of births, will stay constant.

The fear of overpopulation in the years after 1918 and the resulting debate on the population issue were sufficient reason for ‘t Hooft to work out his theory further. He was not impressed with the quality of the existing Dutch literature

Vignette 10: Florus Willem 't Hooft (1896-1941)

When the name of engineer 't Hooft is dropped in the company of elderly Dutch demographers and population forecasters who still remember the generation of their inter- war predecessors, a likely reaction is: 'Ah, 't Hooft, that peculiar engineer, the fellow was a bit of a nut, wasn't he?'.

No reputation of an inter-war population forecaster deteriorated so much as that of 't Hooft after the Second World War. This is surprising because, among the forecasters of the 1930s, his reputation was good. His name was often bracketed with that of Wiebols. 't Hooft's contemporaries definitely did not consider his contribution to population forecasting inferior to that of Wiebols. Leading professionals in the field held a high opinion of 't Hooft. In the second part of the 1920s Verrijn Stuart was eager to con vince the editorial Board of *De Economist* that his contribution on the future growth of the population of the Netherlands should be accepted, because he thought it would provoke a debate on the population problem. And so it did.

Who was F.W. 't Hooft, what was his contribution to population forecasting and by what tricks of fate did he come to stand in such bad repute?
Other than through his writings, not much is known about him. He was a mechanical and a marine engineer. He became interested in the population problem when, at the age of 17, he saw a conveyor belt being tested. The metaphor of the conveyor belt became the core of his theory of the dynamics of population growth. Granules of grain on the conveyor belt represented the inhabitants of a nation and the conveyor belt itself represented the average age at death of the inhabitants. The metaphor became his fate and brought him his fame. The seemingly convincing force of the metaphor blinkered him against other insights and settled his fate. His stubborn adherence to the theory of population growth based on the metaphor resulted in an enduring debate, to which almost the complete intellectual elite of the field contributed. This was the cause of his fall into oblivion after the Second World War. At the same time the metaphor brought 't Hooft fame, because it conveniently conveyed his interest in demography and future population growth. The fierce opposition challenged him to become an expert in his field of interest, forcing him to find other solutions. These marked him as a revolutionary in forecasting methodology and a member of the small group of Dutch innovators of the inter-war period. The debate contributed to the propagation of knowledge of the basics of population dynamics in the Netherlands (De Gans, 1993).

Figure 5.1. 'T Hooft's Conveyor Belt Metaphor

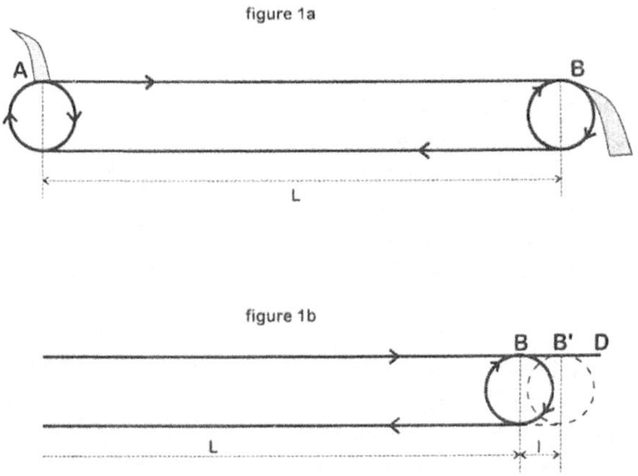

LL	=average span of life of the deceased
l	=increase of the average span of life
A	=the number of births
B; B'	=the number of deceased before and after the increase of the average span of life

on this subject, including the estimates by Methorst (1922), Oly (1924) and Wiebols (1925).[62] His criticism basically amounted to a contention that none of these authors had realized that it was only the continuously increasing span of life which should be cited as a cause of the abnormal increase in the population. He believed that the increase was of a temporary nature; it was just a matter of waiting for the moment when an end would come to the process of the ever-lengthening span of life ('t Hooft, 1926, p. 704).[63]

---

62 The criticism included the forecast of Bowley, which 't Hooft knew from Carr-Saunders book 'Population', as well (London, 1925). Bowley's conclusion that the results of his calculations, namely that the population of Great Britain would reach a maximum in 1971 after which date it would decrease slightly, was "*quite unexpected by most people*", should not have been a surprise at all: "*Apparently professor Bowley was not aware of what is immediately clear from the theory of the conveyor belt, that his results of a population, which has reached the stationary stage after about a lifetime, is the logical consequence of his assumption of a constant number of births and constant mortality conditions; a stationary population number results automatically from it*"('t Hooft, 1929, p. 73).

63 For his works, see 't Hooft (1926, 1927a, 1927b, 1928, 1931, and 1936). Polemics took place with Holwerda, Wiebols, Ubbink and Derksen, stemming respectively from Holwerda (1926) [see 't Hooft (1927a)] and Holwerda (1927) [with Postscript by 't Hooft

't Hooft proposed two different forecasting methods in a row. The first, the conveyor belt approach dating from 1926, was intended as a correction to Methorst's method of 1922, at least as far as the mortality component was concerned. As an adherent to the theory of demographic parallelism, 't Hooft expressed little concern for the future course of the birth rate. In his view the increase in the span of life was the determining factor. The birth rate would follow the death rate. Furthermore, too much literature had already been written on the birth rate.[64]

The method entailed an attempt to correct the future crude mortality rate for the presumed future increase of the span of life. Like many others, he was convinced that an end would come to the population growth in the foreseeable future. At that moment, the birth rate would equal the death rate. It was therefore a matter of determining when the process of prolonging the span of life would reach its zenith and what the level of the (rectified) crude mortality rate would then be. In terms of the conveyor belt theory, it was a matter of ascertaining the moment at which the belt no longer had to be lengthened.

The significance of the declining death rate during the previous decades was not to be underestimated. 't Hooft's effort to standardise the future crude mortality rate was a good idea (like all crude rates, it consists of a population structure component and a mortality component). The real problem, and the core of the controversies 't Hooft opened up, was that he made a correction for the increasing span of life and not for future changes in the age structure. Moreover, the increase in life span was understood not as the development of life expectancy at birth, but as the development of the average age *of the deceased*, calculated annually from the death statistics and extrapolated for the advance calculations. Besides that, he caused misunderstandings by designating his substitute population 'stationary'. Unintentionally, he created

---

(p.p. 153-155)]; Wiebols (1927) [see Postscript by 't Hooft (p.p. 318-319)] and Ubbink (1937) [see 't Hooft (1937)]. Finally a critical contribution in *De Economist* from Derksen (1937) was reacted on with 't Hooft (1937). 't Hooft's book of 1929 included a compilation of his works up to that point. Book reviews appeared in *De Economist* (Van Zanten, 1930), in *Het Verzekerings-Archief* (Oly, 1930) and in *De Socialistische Gids* (Spier, 1930).

64 In a letter, dated 16 July 1926, to the editors of *De Economist* (Archives of *De Economist*, file 1926-1927). 't Hooft was the most outspoken Dutch adherent to the theory of demographic parallelism criticised so severely by Saltet and Falkenburg twenty years previously. It was directly consequential to the forecasting methodology he developed.

the impression that he was referring to the stationary population of stable population theory. It was not until 1937 that he admitted that he should have made better choices in his terminology ('t Hooft, 1937, p. 73).

Even though his conveyor belt theory was not based on stable population theory, his method and hypotheses were *stricto sensu* only valid in stationary situations.[65] Another weakness was his disregard of fertility as an independent factor, in particular its influence on the age structure with all its consequences for future numbers of births and deaths and the crude rates.

5.7 | ... and 't Hooft's Innovative Cohort Approach

Because of its internal weaknesses, 't Hooft's conveyor belt theory can definitely not be considered as an improvement to forecasting methodology. The method introduced a year later, in 1927, fared better. It was a cohort approach, based on the reconstruction of the quantitative course of development of ten-year birth cohorts, not differentiated by sex, from the data of censuses held from 1830 to 1920 in ten-year intervals (*Figure 5.2*).

Starting with the birth cohort 1820-1829, 't Hooft calculated how many of a hundred initially 0-9 year olds were, after 10, 20, 30 years, still alive and in the Netherlands. Each ten-year cohort was analysed to determine how diminishing by death took place. Although as far as is known the method was developed independently, it was very similar to Edwin Cannan's forecast of 1895. Cohort-specific 'mortality and international emigration surplus' proportions ('survival in the Netherlands' proportions) were calculated and extrapolated. Just as Methorst, Oly and Wiebols and many other forecasters before and after him, 't Hooft felt that foreign migration could be omitted. On the basis of the development of the figures in the successive birth cohorts in the observation period 1830-1920, 't Hooft made a forecast of the future development of each ten year birth cohort and age group by a smooth extrapolation of the declining cohort lines and the rising 'similar age class' lines he found ('t Hooft, 1927b). With respect to the decrease in the figures of the coming

---

65 That was in itself no problem, because in the case of changing mortality conditions, *semi stable or quasi stable* situations would have come about, whereby part of 't Hooft's theory would still remain standing (Pressat, 1983, p. 258 et seq.).

Figure 5.2. 'T Hooft's Diagram. Cohort 'Survival-in-the-Netherlands scheme, 1830-1920 (observed); 1920-1980 (extrapolated)

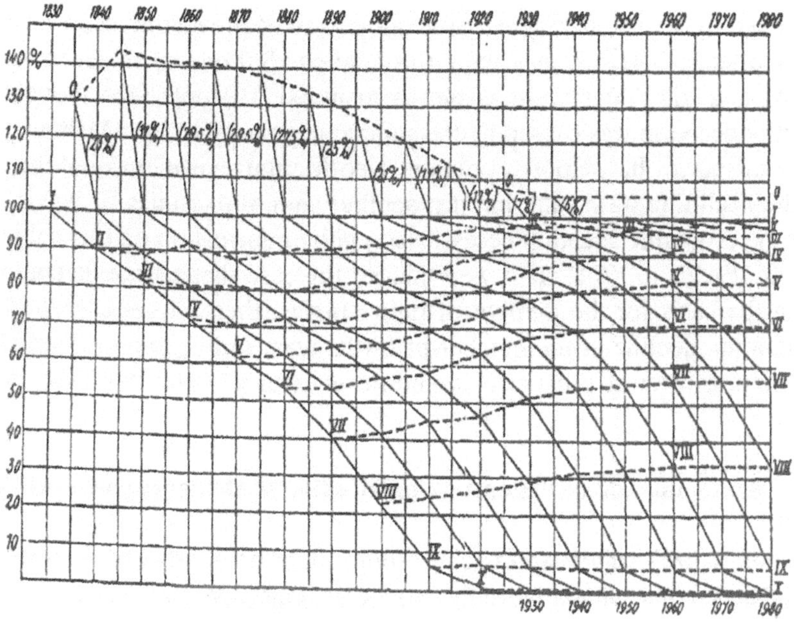

Source: 't Hooft, (1927/1929).

birth cohorts, he relied on extrapolation of calculated past ratios of the observed number present in the age class 0-9 years in the successive censuses and the number of live births in the ten-year interval immediately previous to the present one.

A separate forecast was made of the future number of live births. Given his demographic parallelist inclination, it is self evident that he assumed there would be a continuing drop in future births, at the current rate to start with, then decelerating until a balance between births and deaths was reached. 't Hooft was opposed to those who believed that fertility decline could be explained from progress in civilisation and increasing hedonism, or from Neo-Malthusian propaganda. Rather than adhere to social Darwinian beliefs, he was convinced that the single major cause of fertility decline was socio-economic: The economic conditions, accompanied by the standard of living to which people became accustomed. In his view, no-one wishes a lower rank on the social ladder for himself or his children. Marriages only take place when maintenance of the standard of living to which one has become accustomed seems likely to be assured. Whether limitation of family size be achieved by sexual abstinence or in some way of preventing conception did not matter to 't Hooft, at least in this context. But one must remember that attempts at the latter were as old as mankind.[66] People who thought that fertility decrease was caused by the gaining popularity of Neo-Malthusianism were mistaken. In 't Hooft's opinion, Neo-Malthusianism was not the cause of the declining birth rate, but just another sign of the times. It was evident that the economic situation necessitated public information on this subject ('t Hooft, 1927b, p. 671). If it were indeed the case that economic conditions form the basis for the number of future births, then a forecast involving a continuing decline in the number of births was, in his opinion, completely justified, as prewar prosperity would never return ('t Hooft, 1927b; 1929).

5.8 | An Evaluation of 't Hooft's Contribution to Demographic Forecasting

't Hooft's 1927 cohort approach to future population offered an insightful, useful and simple alternative to the more complicated Wiebols method of 1925. The Wiebols method was more advanced, however, because the age structure formed the explicit basis of the calculations. Future births were linked to the changing age structure by way of the general fertility rate. Moreover, as it turned out, migration could be included in the calculations with a minimum of effort.[67] Inserting age specific fertility would have been possible with a sex-differentiated cohort approach too, but it would be impossible to insert migration. In 't Hooft's model, emigration, immigration and survival were bundled into a single 'survival in the Netherlands' rate and they could therefore not be differentiated from each other.

Although 't Hooft distanced himself from 'statisticians' like Oly and Wiebols who in his eyes were only equipped to extrapolate current situations into the future, he did the same thing himself. He took pride in the fact that he, in contrast with the 'experts', relied on the practical situation: He used data

---

66 't Hooft was decidedly not alone in these interpretations. See for instance Bierens de Haan (1924).

67 Formal cohort component modeling would reach its apex in Leslie's matrix model, developed in the 1940s.

gleaned directly from the censuses. The fact that he, too, worked with the probabilities which he himself criticised, even if they were not pure survival probabilities, escaped him.

Nevertheless, his method was innovative and functional for the Netherlands at a time when population forecasting was developing swiftly and insight into the background of the processes of population change was sparse. He was able to offer a clear insight into the changes in the totals of birth generations through the years, particularly through the graph illustrating his method, and the resulting changes in the age composition of the population.

'‘t Hooft's 'cohort survival in the Netherlands' methodology comes close to generation life table methodology, of which he was a convinced supporter. In its application he was far ahead of his time, as some of his contemporaries recognised. Van Zanten for instance stated that the Wiebols method was more refined than that of 't Hooft, but that of probabilities of death of a then current population group were used for the distant future (a period mortality table of civil servants). In contrast, the purport of the 't Hooft method was that it was necessary to work with generation life tables (Van Zanten, 1930, pp. 319-310; also CBS, 1975, p. 10). 't Hooft put into practice what the Dutch life table expert Van Pesch had already called attention to many years before (Van Pesch, 1912; 't Hooft, 1929, p. 96). 't Hooft kept pressing on a weak spot in the approaches of Oly and Wiebols: Their use of period life tables.

An attendant result of his activities in the field of generation mortality was that 't Hooft, as far as is known, was the first to introduce the 'baby boom' concept (which he called the 'tidal wave'; 't Hooft, 1927, p. 676). No-one else in the Netherlands described the consequences of the great number of births in the years following World War I and its effects on the population over time so expressively, nor tied the changes so closely between the need for amenities and services to changes in the age structure, themselves a consequence of the 'advancing' effect of consecutive birth cohorts. In the imaginative evocation of things to come he stood close to Westergaard's visionary horoscope of the population in the 20th century. At about the same time that Warren S. Thompson formulated the relationship between level of fertility and social economic factors in a comprehensive way, 't Hooft formulated the core concept of demographic transition in a more concise, imaginative way than it ever had been done before in the Netherlands. It is also one of the few instances in which 't Hooft appeared to be able to distance himself to some extent from his mechanistic conveyor belt theory.

5.9 | Why and by Whom 't Hooft was Maligned...

In the Netherlands the reception of 't Hooft's work was controversial. His contribution was both severely maligned and highly praised. From the very first article in which his conveyor belt theory was put forward (in 1926), reactions were exceedingly negative. Holwerda (1935) spoke of a regrettable decline in quality after Oly and especially after Wiebols. Presumably it was De Jong, who had the most detrimental effect on his post-war reputation. Following Holwerda and colleagues, De Jong suggested that 't Hooft's work could be disregarded in a discussion of the reproduction intensity of a population because it was "*if not worthless, then merely partially correct and in any case impractical and ineffective*" (De Jong, 1946, p. 164 f.n.2). Van Praag (1976) spoke of an interruption in the rising progressive line of the development of forecasting science.

The dismissal of 't Hooft's work as a serious contribution to the innovation of forecasting methodology cannot be completely explained by the rejection of the conveyor belt approach, although he managed to gather attacks from quite a few experts in the field of statistics, forecasting science and demographic theory.

The editorial board of *De Economist* had looked for an intensive debate when his first article was published in 1926. Verrijn Stuart thought that the importance of the issue justified continuation of the debate he had started in 1922. Verrijn Stuart expected polemic reactions, but even he may have been surprised by the vehemence of the discussion 't Hooft's article provoked.[68]

It is not clear why 't Hooft continued to uphold his conveyor belt approach with such exasperating obstinacy in the face of such consistently sound criticism. His correspondence with the editorial staff of *De Economist* and his publications show he was well informed of the latest advances in analytic demographics, especially on matters concerning mortality, mortality decrease

---

68 Letter, dated 8-9-1926, from editorial board secretary G.M. Verrijn Stuart, C.A. Verrijn Stuart's son, to the members of the board (Archives of *De Economist*, file 1926-1928).

and the influence of the prolongation of the life span on the growth of population. He discussed drafts of his articles with experts. Since Methorst, whose views were adjacent to the theory of parallelism, was one of these, it is conceivable that part of 't Hooft's obstinacy came from the fact that he thought he had an ally in Methorst, though the evidence for this presumption is very circumstantial.[69] It is more likely that a mixture of causes was at work.

One of these causes, and not the least important, was 't Hooft's absolute conviction that the main cause of population increase was the prolonging of the average age at death, and that this was a temporary phenomenon. Another was that he became obsessed by his conveyor belt metaphor and lost his sense of self-criticism. He may have suffered from the tunnel vision of an amateur in the field of experts who finds maintaining the status of the outsider a matter of honour. To start with, he had no high opinion of the existing literature concerning the population issue ('t Hooft, 1926, p. 47, 1927a). In this he shared Wolfe and Bouthoul's opinions of the international literature (see Chapter 4). However, in contrast with 't Hooft, Wolfe was able to distinguish between the general literature of contributors to the debate on the population issue, for which he held no high opinion, and the contributions of such innovative forecasters as Cannan and Bowley, which he valued. In any event, 't Hooft tended to explain the lack of understanding he encountered as resulting either from his position as an outsider, or as an innovator with a new theory against traditional professionals who were reluctant to accept new ideas ('t Hooft, 1927a, pp. 55-57), or as a practical man working with real data ('t Hooft, 1927a), or as an engineer who had tackled the population issue as a problem of physical science ('t Hooft, 1927c, p. 319, 1937, pp. 370-371).

His perceptions of opponents and the soundness of their criticisms might have been reinforced by the way Holwerda and Wiebols treated him. He felt they had snubbed him. Holwerda's criticism must have hit him especially hard: He had discussed an earlier version of his 1926 article with both Holwerda and Methorst. 't Hooft was convinced that he had accommodated the criticism Holwerda had expressed in that earlier meeting.[70] With respect to Wiebols,

---

69 It is striking that Methorst never took part in the debate, or took sides. Neither did he ever –as far as is known– refer to 't Hooft or to the discussion mentioned. That he kept aloof can be explained as unwillingness to burn his fingers in a matter that could harm his reputation.

70 Letter, dated 19 June 1926 (Archives of *De Economist*, file 1926-1928).

the editors of *De Economist* made the mistake of relaying a criticism of his theory which 't Hooft felt was phrased extremely ungraciously.[71]

For Holwerda, De Jong and Van Praag, the criticism of 't Hooft's conveyor belt approach was sufficient reason to malign the man's entire contribution. It was detrimental to his post war reputation as a forecaster. In fact, his critics were so successful that the image left in the minds of post World War II demographers was that of a maverick, despite the positive reception to his cohort approach to forecasting in the thirties and the mid-forties.

5.10 | ... and by Whom 't Hooft's Work was Applied

't Hooft's image as a forecaster may have finally taken on a negative colour, but that does not alter the fact that many of his contemporaries greeted his contribution to forecasting positively almost immediately. His cohort approach was generally ranked with the Wiebols method. Oly (1930) was one of those who, despite criticism of the conveyor belt metaphor, had great esteem for 't Hooft's work. Others, such as the statistician Derksen, thought that the issues were so important, they could not remain unaddressed. Derksen was also of the opinion that the criticism had sometimes been too harsh (Derksen, 1937, p. 281). 't Hooft's cohort approach was ranked with that of Wiebols (Verrijn Stuart, 1927, p. 69 f.n. 2; Van Zanten, 1930, p. 319, 1938, p. 155). That 't Hooft's approach was taken seriously is also clear from a discussion of forecasting in a study by Van Lohuizen and Delfgaauw of the future expansion of the population of Amsterdam foreseen in the Amsterdam General Extension Plan of 1935 (Grondslagen, 1932, p. 104-106; also Bakker Schut, 1933; Angenot, 1934). As late as 1942 't Hooft's approach was used in the demographic foundations of the Regional Plan of Eindhoven and Environs (*Streekplan* ... 1942).

Towards the end of the 1930s, the conveyor belt forecast played a part in a new phase in the debate on the population issue, then centred on the threat of a population decrease as a result of the diminishing fertility rate. Van Vuuren, a professor of social geography in the University of Utrecht and one of those who viewed voluntary birth control as a danger to the nation, based a pamphlet

---

71 Correspondence dated 10 March 1927 and 13 March 1927 (Archives of *De Economist*, file 1926-1928, under 't Hooft).

warning the Dutch nation of the dangers of fertility decrease on 't Hooft's calculations. There appeared in *De Economist* a serious reprimand from Ubbink, a fellow engineer who was, to 't Hooft's disappointment, nevertheless one of his critics (Van Vuuren, 1936, 1937, p. 376; Ubbink 1937).[72]

Significance

What is 't Hooft's rightful place in the intellectual history of modern population forecasting in the Netherlands and what is the true significance of his contributions? Many of his contemporaries considered his cohort approach an alternative of equal standing to Wiebols' forecast. In the years following the publication of 't Hooft's collected works in 1929, three municipal teams published studies: Van Lohuizen and Delfgaauw in Amsterdam (see Chapter 1), (Grondslagen, 1932); Bakker Schut, director of The Hague Department of Urban Development and Housing (1933) and Angenot, engineer/town planner at the Rotterdam Department of Urban Development (1934). Each of these studies compared the merits of Wiebols' method and 't Hooft's cohort approach.

In the Amsterdam study, the mortality scheme used by 't Hooft to calculate the future mortality rate was judged to be essentially the same as Wiebols' method, although the latter was deemed more refined in terms of probabilities of death for the first years of life.[73] The final choice in favour of Wiebols' method resulted from the fact that Wiebols used the general fertility rate and not the much simpler crude birth rate, as 't Hooft had done (Grondslagen, 1932, p. 105-106).

The 't Hooft methodology played an important role in the forecasts of Bakker Schut and Angenot. Whereas Bakker Schut deemed the use of fertility rates by Wiebols to be principally correct, he felt that the factor 'fertility' was too sensitive and depended on too many subtle influences to be capable of indicating a future course of events with a high degree of probability. He preferred 't Hooft's cohort approach because of his use of mortality schemes in ten-

---

72 Ubbink (born 1902) was not very interested in demography, population forecasting, or the population issue; neither did he know much about the subject. As a mechanical engineer and reader of *De Economist*, he felt that something was wrong in 't Hooft's argumentation and he could not help reacting to it (private communication to the author).

73 From a technical standpoint, this is not the case. In 't Hooft's cohort method it was impossible, for example, to insert migration as a separate factor into the calculations. The future number of births was also estimated exogenously.

yearly inter census periods. However, Bakker Schut himself preferred to work with the conveyor belt approach because in his opinion this was connected to 'more objective factors'; the average length of life and the absolute number of births per year.

It is interesting to note that Van Zanten approved Bakker Schut's preference for 't Hooft over that of his own Amsterdam colleagues Van Lohuizen and Delfgaauw's based on Wiebols' method. This was because, according to Van Zanten, Bakker Schut was interested in calculations for a practical situation, namely the need for amenities and services in the Hague, and not in a scientific demographic study. Roundabout, laborious calculations as in the Amsterdam method might evoke trust, but they remain nonetheless speculative (Van Zanten, 1933a, p. 591). Angenot, too, relied on the work of 't Hooft. In his view, 't Hooft, along with Oly and Wiebols, belonged to the group of professionals who had helped advance correct insights into the matter of the future course of the population (Angenot, 1934). He felt the methods of Wiebols and 't Hooft both had their strong points and were essentially the same. By combining them, Angenot attempted to improve them both. He converted the extrapolated age-specific death rates employed by Wiebols (and which he did not consider needed improvement) into generation cohort survival schemes as 't Hooft had done. So Angenot did not work with generation life tables as 't Hooft had propagated. He merely converted the probabilities of death used by Wiebols, adjusted for Rotterdam, into survival schemes. In Chapter 7 the merits of the forecasts of Amsterdam, The Hague and Rotterdam are discussed from the perspective of municipal forecasting.

The Sting of Criticism

It is worthy of note that the positive reception of 't Hooft's work by municipal forecasters/town planners of Amsterdam, The Hague and Rotterdam was not reflected in the debates of the 1930s. It is equally noteworthy that no references to the weak points of 't Hooft's methods can be found in the publications of Bakker Schut or Angenot, nor in Delfgaauw's reviews of their forecasts (Delfgaauw, 1933b, 1935). This is surprising in view of Bakker Schuts' ready employment of the conveyor belt approach, Angenot's building on the cohort approach and Van Lohuizen and Delfgaauw's positive summary of both contributions in their methodological introduction to the Amsterdam forecast. With respect to 't Hooft's methodological merits, there seems to be an impenetrable line of demarcation drawn between participants in the debate on national population forecasting methodology and municipal town planning engineers combining an interest in population growth for town planning purposes with an interest in national population development.

Closer inspection, however, shows that the town planners removed the sting of criticism from the theory on which 't Hooft had based his work. Bakker Schut did so by (re)defining 'average span of life' not as the average age at death of the deceased of a given year, but as the life expectancy of that year's new-borns; Angenot by limiting himself to the discussion of the influence of the increase in the span of life on the mortality rate in a stationary population. Both, therefore, violated the essence of the conveyor belt theory. Whether there lay at the root of their criticism a basic misunderstanding of his theory in interpreting it in terms of stable population theory, or whether there was a conscious removal of the sting of criticism from the surveyor belt theory cannot as yet be ascertained. In the event, his basic theory was applied within that one instance where it was indeed correct, that of a stationary population (instable population theory). In this respect Bakker Schut and Angenot were in agreement with the earlier assumptions of Oly and Methorst, that a stationary situation would be reached in the foreseeable future. They deemed it possible, if not probable, that the population of the Netherlands would finally become stationary (Bakker Schut, 1933).

5.11 | A Contemporary State-of-the-Art Overview

How the cohort approach, in terms of 't Hooft's survival-in-the-Netherlands proportions, was put to optimum use is discussed in the Chapter 7. An excellent summary of modern population forecasting methodology at the end of the inter-war period was made by Angenot in an overview accompanying the publication of the NCBS forecast of 1951. Angenot feared that outsiders merely glancing at the list might receive a confused impression. Sometimes, for instance, two forecasts with different results had been published in the same year. The apparently capricious selection of methods and assumptions might lead an outsider to think that forecasting was seen as a pastime. Such an impression would be totally untrue. There is a positive development to be noted in the Netherlands. Angenot thought it sensible to distinguish between domestic and foreign forecasts. In his opinion the domestic forecasts were much better than the foreign ones. Wiebols was the pioneer, in departing from age-structure. Doing so he had made the methods of Verrijn Stuart, Methorst and Oly obsolete. Most later authors followed Wiebols, but with methodological improvements. 't Hooft introduced a generally accepted technological improvement by using survivor-percentages. "*I myself introduced a refinement by using age specific female natality risks, which were calculated by me for the first time. From a methodological point of view the calculation methods of Bakker Schut and Reitsma stand apart. Their methods did not find any imitation. It must be remarked that in all prewar assumption making calculations were based on a zero migration surplus.*" (NAi, Arch. Angenot, inv. Nr. 73).[74]

5.12 | Conclusions

The development of population forecasting in the Netherlands came about in a brief time span. After 1922, any calculation of the future population of the Netherlands which was to be taken seriously dismissed the geometric method. From then on, the extent of the future population was calculated on the basis of analysis and forecasting of the components of natural population growth. The insight that the age structure was of paramount importance, even if one was only interested in the course of the total population, gained strength. After 1925, any serious calculation had to start from the age structure of the population. In 1937 it received the hallmark of approval by the greatest authority in the field of statistics and the investigation of population: Methorst made clear once and for all that age structure was 'the pivot of the population issue' (Methorst, 1937). As was his style, he felt no obligation to refer to the earlier contributions of his compatriots.

As demonstrated by the calculations by Rooy and Verrijn Stuart, the geometric method was not applied indiscriminately in the Netherlands. In assumption making the future development of the vital population components of growth, natality and mortality were taken into consideration in determining the future growth rate. The idea of a 'law' of geometric population growth disappeared. The method of logistic population growth never became part of mainstream forecasting methodology. Only in an isolated, rare case was it taken as a 'law'.

The inter-war period was characterised by a struggle for the 'best' method, to be compared with the international debate on the merits of respectively

[74] Reitsma's forecast was one of the few Dutch inter-bellum forecasts that started from logistic growth methodology (Reitsma, 1935).

logistic growth methodology and the demographic approach. But the contention between the methodologies of Wiebols and 'Hooft was a struggle within the context of demographic analysis. It took place across successive publishing years of the oldest economic journal in the Netherlands, *De Economist*; which can be taken as a sign of the importance attached to this issue.

De Economist's coverage was definitely a sign of Verrijn Stuart's involvement in the issue. He paved the way for the beginning and took care of the continuation of the national debate on the population issue. His view of statistics withheld him from involving himself in the actual population forecasting business and kept him from a true interest in methodological forecasting issues. He agreed with 't Hooft's view of the future course of natality. By allowing 't Hooft ample space in *De Economist,* Verrijn Stuart served his own purposes as well. Being a stubborn man himself, he may even have felt some congeniality of character with 't Hooft.

It is not improbable that the sympathies of the international representatives of statistics, Verrijn Stuart, Methorst and Van Zanten tended to side with 't Hooft rather than with Wiebols, albeit for different reasons. Of course Methorst never sided with either of them in public. He was not a man who easily took sides; even less to risk his reputation. In how far he was instrumental in sustaining, or even encouraging, 't Hooft's views of demographic parallelism is a matter of speculation. In any case when 't Hooft sought his counsel he failed to convince him that his conveyor belt approach had led him onto the wrong track.

In Van Zanten's view, population forecasting had to be as simple as possible if it were to be applied to practical situations. Sophistication was to be reserved for scientific purposes, because it suggests unwarranted accuracy. Given his private view of population forecasting it is understandable that he preferred Bakker Schut's application of 't Hooft's conveyor belt theory to the more sophisticated method of Wiebols and his colleagues of the town planning department of Amsterdam.

The true methodological innovation of population forecasting came from mathematical statistics in the persons of Holwerda and Wiebols, and from 't Hooft, an interested and pragmatic outsider who did not shy from mathematics or quantification and who was not easily impressed by the status of authorities in the field. Wiebols introduced age-structure and age specific fertility rates into forecasting, ‘t Hooft advocated the cohort approach and gave the first, most vivid Dutch picture of demographic transition and its consequences. It is also to the credit of ‘t Hooft that he succeeded in provoking his opponents into a debate on the method of population forecasting. In this way he contributed to the dissemination of knowledge and insights into the analytical-demographic backgrounds of population dynamics and forecasting. His graph of the ‘survival in the Netherlands’ proportions in successive birth cohorts was extremely enlightening and possibly the first of its type in the Netherlands, comparable with but independent of that of Cannan in 1895. The cohort approach introduced by ‘t Hooft (actually a ‘still present in the Netherlands’ approach on a cohort basis) found hardly any further application. Generation life tables never gained a foothold in population forecasting (Keilman, 1990). Lastly no one in the Netherlands of the inter-war period gave such a vivid and enlightening picture of population dynamics and its workings as did ‘t Hooft with his metaphor of the ‘tidal wave’ of births and the course of the wave through the age-structure and its consequences. Its evocative power is reminiscent of Westergaard’s horoscope of the population of Europe.

6. Forecasting Future Housing Need in the Netherlands

6.1 | The Impact of the Housing Act (1901)

The context of urban population forecasting was totally different from that of national forecasting. At the municipal level practical utility stood central, not the scientific aspect of forecasting. Forecasters had not only to find an answer to the challenge posed by migration and the arbitrariness of municipal boundaries; they had also to face the challenge of estimating the number of households and forecasting future housing demand.

The Housing Act of 1901 can be considered as the formal beginning of urban and regional population forecasting and physical planning in the Netherlands. The Act also became the legal basis of official housing policy. The authority was given to municipalities to raise housing conditions. For that reason, municipalities needed a better insight into the kind and extent of housing demand and the development of housing need. This was the core of the Housing Act (Van der Schaar, 1991, p. 88). It was to regulate the environmental development of town and country for a long period of time. It stood at the cradle of housing need estimation, discussed in the present chapter, and local and regional population forecasting for planning purposes, discussed in Chapter 7. The Housing Act established a close link between population forecasting, housing and town planning.

The effect on the development of physical planning and municipal population forecasting is comparable with the effect of the debate on the population problem initiated by the Society of Political Economy and Statistics (VSS) on the innovation of national population forecasting. The Housing Act furnished the regulations dwellings and their use had to meet. It obliged all municipalities of the Netherlands to assess and verify building regulations. Municipalities with a population of twenty thousand inhabitants or more were obliged to set up a housing registration system and keep it up to date. The Housing Act allowed the national government to lend money to municipalities for social housing purposes (Smit, 1995).

The Housing Act became operational in 1902, together with the Health Act. Social housing projects could be undertaken on a much larger scale than the private initiatives of the 19th century. Municipalities of ten thousand inhabitants or more and municipalities which had seen a population growth of 20 per cent or more in the past five years were obliged to make an extension plan (De Gans, 1989). In the big municipalities in particular, often quite powerful municipal housing departments were established, as for instance in Amsterdam in 1915. In this city the Municipal Housing Department was to operate as a housing corporation for the lower income groups. The Department did not limit itself to the actual building of dwellings; it was involved in all policy aspects of housing, town planning included. As to the future development of Amsterdam, the Housing Department and the Department of Public Works were fierce opponents in the policy issue of opting for the construction of garden cities or town extensions.

6.2 | World War I and the Housing Shortage Issue

Prior to World War II planning was for a great part the mere adjunct of housing. The estimation of the (future) number of households and of (future) housing need was one of the big issues, for instance for the foundation of municipal extension plans. Notwithstanding, it took quite a long time before demographic forecasting for housing purposes came into focus. As long as future housing demand was calculated from average family size (or average household size), town planners could suffice with traditional forecasts of total population. In the course of time the defects of this approach were exposed, without however ever resulting in its complete disappearance.

Under the pressure for accuracy in estimating current housing need and current housing shortage it had already become clear before the First World War that there were better methods of making such estimations. For the time being these methods did not stimulate the application of demographic forecasting innovations at urban and regional levels, but they were at the very basis of the development of family (and household) demographics which gained international and national momentum from the 1970s onwards.

Apart from the interest in the accurate estimation of current and future housing need, the focus was on dwelling type distribution, because town planners were interested in the kind of housing districts that municipal extension plans had to provide for. The focus of interest was the population composition in terms of social economic or income categories, not age structure. It was only in the first part of the 1930s that the advantage of age (and sex) specific population forecasts for calculating future housing need became apparent.

The first years after the end of the First World War were characterised by the issues of the size of the housing shortage and the estimation of future housing demand, attributed to the lack of investments in housing in the war period. Estimates of total housing need in the period 1918-1923 ranged between 250,000 and 300,000 houses (Bosma, 1993). These totals agree with Rooy's estimation of necessary average building production of about 50,000 per year, as discussed in Chapter 5. The size of the problem had increased considerably in comparison with pre-war years. In terms of its impact on the building of subsidised dwellings, the immediate influence of the Housing Act in the first decades of the century must not be overrated. Between 1918-1923 the Government spent yearly about twenty times the amount of money to subsidised housing as in the year 1914 (Bosma, 1993).

Estimation of municipal housing need was one of the big challenges municipalities had to face. After World War I in particular, housing statisticians had to find ways to make sound calculations of the number of dwellings to be built to address the problem of current housing shortage. In general, attention was focussed less on the calculation of future housing need than on the estimation of current housing shortage and current housing need. This was difficult enough because of the lack of sufficient statistical data (e.g. Bakker Schut, 1915a; Rikkert, 1919).

According to Faber, the ideal way to estimate current housing need was to start from the three factors causing it: The formation of new families, the loss of existing dwellings, and the immigration surplus. Because of the difficulty of acquiring a good insight into the factor of the formation of new families, Faber calculated a proxy instead. He estimated the housing need by dividing total population increase by a dynamic average family size figure. Continuation of the previous use of a constant average family size figure of five persons per dwelling would result in an underestimation of the true housing need, because the figure had already decreased from 4.54 in 1899 to 4.47 in 1909 (Faber, 1918).[75] In principle, the method could be used for estimation and forecasting purposes. In the latter case, two separate forecasts were needed: A forecast of future total population, and an extrapolation of the average family size figure, average household size figure or average dwelling occupation figure. The method is simple and pragmatic and is still in use.

The calculations of Rooy, discussed in Chapter 5, and Faber give an indication of the state of the art of housing need forecasting at the beginning of the 1920s. It is possible to assess the state of the art at the end of the decade on the base of a methodological overview in the report of the Amsterdam Garden City Commission (Rapport, 1929). The Commission's assignment was to investigate the necessity of the foundation of new garden cities in the neighbourhood of Amsterdam. A subcommittee of the Garden City Commission was asked to calculate what increase in the Amsterdam population was to be expected in the near future and the necessary town extension that would result from it, while taking into account the demands posed by the improvement of housing (Rapport, 1929). Meanwhile, as explained in Chapter 5, national population forecasting methodology had experienced innovative stimuli from Methorst, Oly, Wiebols and 't Hooft, while the merits of the methods of Wiebols and 't Hooft were being debated vehemently in *De Economist*.

The committee's introductory comments of the calculation results are promising, because they echo the innovation of national forecasting methodology. The committee started by indicating that one of the options considered for the calculations had been an analysis of the factors that determine population increase, followed by a separate extrapolation of the observed trend line of each of these factors and the addition into a final total for the year to be forecast. This would mean that total population change had to be split up into a natural population increase factor and a migration surplus factor. Natural

---

[75] Faber wanted to make up for the omission that no estimation of the size of the current housing shortage was made for the first National Housing Conference of 1918, which was organized by the National Housing Council.

future population growth, it was stated, could be calculated in different ways, for instance by linear extrapolation of the natural population increase, or by splitting up natality and mortality, followed by separate extrapolation of both graphs and combining them. Future natality could also be calculated from estimations of future fertility and the future age structure of the population, while for mortality the future age structure and mortality rates could be examined as separate components. Even further analysis was conceivable (Rapport, 1929, p. 75).

One is inclined to see the reflection of some of Wiebols' remarks in his letter to Van Zanten (see Chapter 2). It is likely that Van Zanten's Amsterdam Bureau of Statistics was asked to provide the data for the calculations. The person responsible for the calculations of the subcommittee was professor H. Frijda, an economist at the University of Amsterdam who was acquainted with Van Zanten. A few years before Frijda had recommended Van Zanten to the Faculty as a lecturer in statistics (De Gans, 1994a).

Promising as the above mentioned introductory remarks to the report of the Garden City Commission may appear, they lead to nothing if the report is evaluated in terms of its contribution to the innovation of forecasting methodology. It was argued that *'all such polished calculation methods'* have in common that expert statisticians and mathematicians differ seriously in their opinion of the method to be followed. The probable error would presumably not be less than in the case of an extrapolation of total population increase, because more factors would act as independent causes of error and because migration would be an uncertain element. Lastly, the committee was of the opinion that each of these investigations would be extensive and would require a scientific study of many months (Rapport, 1929). Here again, a glimpse can be seen of Van Zanten's pragmatic vision of population forecasting for everyday planning practice purposes.

Whatever the truth of it, calculating future population growth by simple extrapolation was thought to be good enough. The committee distinguished itself in an assiduous, pragmatic rather than innovative approach to future housing need. Calculations ran from extrapolations of observed absolute increase in the number of families to figures based on expected future total population and expected future average family size; from computations based on future average household size to combinations of the variants listed. In total, seventeen different types of calculations were made, all of a familiar unsophisticated type.

6.3 | The Halle Method

The main innovation of the 1920s was in the estimation of the current number of households, not in the prediction of future household need based on population forecasts. Once the number was known, the number of households in need of a dwelling could be calculated and the housing shortage estimated. The propagation of the new estimation methodology gave rise to a fierce debate among housing statisticians and housing experts.

In 1914 Tellegen, Director of the Department of Building and Housing Supervision of Amsterdam and one of the few housing reformers of the country with active experience in the field, introduced a more satisfactory method than that of Faber. A model of the process of family formation formed the start. The basis was the calculation of the yearly changes in the number of families, which was used as an indication of changes in housing need.[76] The German town of Halle am Saale, where the method was devised and successfully applied, gave it its name: The *Halle method*.[77]

By modern standards, the method was quite advanced; the process of family formation was essentially modelled in terms of marital state transitions. Only many years after World War II did the dynamic approach of family and household modelling and forecasting start to develop along similar lines.[78]

The application of the method depends to a high degree on reliable, up to date statistics of marital state transitions, so it is understandable that Van Zanten

---

[76] J.W.C. Tellegen (1859-1921) had studied civil engineering in Delft. From its establishment in 1901 Tellegen was the Director of the Department of Building and Housing Supervision of Amsterdam. In 1915 he became the mayor of Amsterdam.

[77] It was applied for the first time on December 1, 1905. The results were published in a publication of the Statistical Office of Halle *Die Leerwohnungen in Halle a.S. 1909-1911* (Heft 17, 1912. Halle: Gebauer-Schwetsche). The statistician of Halle was greatly surprised to learn of the renown of the method in the Netherlands (Van Zanten, 1925, p. 145).

[78] See, for instance, De Vos and Palloni (1989); Burch, in: Van Imhoff, *et al.*, (1995); Kuijsten (1996).

had a professional interest in the Halle method and in the 1920s took an active part in the debate on the merits of the method. A concise summary of the method can be found in his manual.

Van Zanten's summary runs as follows. The Statistical Office in Halle investigated the conditions leading to the formation of new families, who would then need a new dwelling and the disappearance of families, which would clear inhabited dwellings. Generally speaking, immigration leads to the former and emigration and death to the latter, although not every wedding results in an increase of housing need, nor does death inevitably result in the clearance of a dwelling. The Office concluded that housing need increases with: The marriage of unmarried people; the marriage of a divorced man with an unmarried woman; the immigration of a family. Housing need decreases with the marriage of a widower and a widow; the marriage of a widower and a divorced woman; the death of a widower, widow or divorced woman; by the emigration of a family. Because the method was not totally accurate, checks and verifications with census data were necessary from time to time. Several factors lead to inaccuracies: Not all marriages of unmarried persons have an increasing effect on the housing need; not all persons marrying a person from outside the municipality compete on the municipal housing market; not all widows or widowers continue to occupy a dwelling after their partner's death. Moreover, cohabitation outside wedlock (concubinage, brothers and sisters or even non related persons living together) is a totally uncertain factor (Van Zanten, 1938, pp. 356-358).

The Halle method allowed for an analysis of the process of family (household) formation and dissolution on a yearly basis and gave an insight into the 'true' development of housing need. The method could also be applied retrospectively by calculating the development of the number of dwelling needing households over the past period. With the Halle method, Rikkert, the Housing statistician of the Department of Housing, was able to demonstrate that the Amsterdam housing shortage over the period 1909-1918 had increased considerably; quite a different conclusion from what could be assumed from the course of development of the average dwelling occupation figure over the same period (see *Figure 6.1*). The shortfall was an important discovery. Rikkert made it clear that the average dwelling occupation figure depended on changes in the demographic factors of family formation and dissolution onthe one hand and changes of the housing stock on the other. Moreover, the extrapolation of observed time series of the average dwelling occupation fig-

Figure 6.1. Average dwelling occupation rate of the Netherlands, 1908-1912.

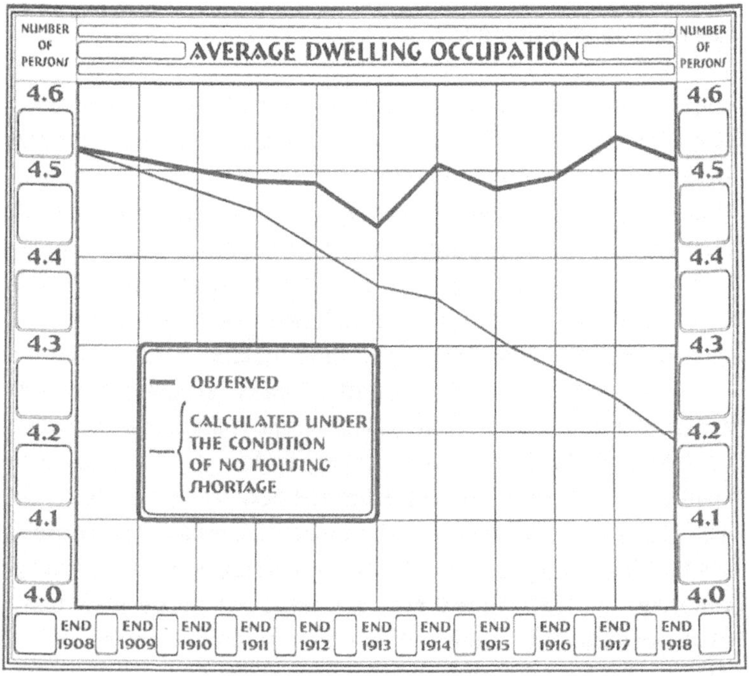

Source: Rikkert, 1919

ure for forecasting purposes was based on unsound reasoning. Rikkert's argument, based on the application of the Halle method, was one of the earliest warnings against the use of the average dwelling occupation figure as an independent instrument for the prediction of future housing need (Rikkert, 1919; Van Fulpen, 1985).

From the moment it was introduced, the Halle method was successfully applied in Amsterdam. People like Rikkert had a high opinion of its accuracy; he became one of its most ardent advocates (Rikkert, 1919; Engberts, 1973). Presumably the fervent advocacy of the housing experts of Amsterdam and other municipalities brought severe opposition, too (Engberts, 1973).[79] In

[79] Schutte of the municipality of Utrecht, for instance, wrote an article on the Halle method in the *Weekblad voor Gemeentebelangen* (Weekly for Municipal Interests) with the sole purpose of making the method more widely known (Schutte, 1922).

1925, the Chief Inspector of Public Health (and Housing) Van der Kaa launched an attack on the utility of the Halle method, asserting that its application had resulted in an overestimation of the housing shortage. Uncritical application of the method could result in undesirable counter effects, as was already the case. Several municipalities decided to renounce any further application.[80]

Wide propagation of the method and political interest in the housing issue in the big cities in the inter-war period led to a debate on its merits throughout the 1920s.[81] The criticisms of this statistical instrument caused considerable unrest. The controversy was largely among experts in the field of housing and housing statistics with direct implications for housing policy, not academics. The debate is a clarifying indication of the social significance attributed to the housing issue.

In 1925, the NIVS Board convened a committee to investigate the Halle method and report on improvements that could be made. Hudig chaired the committee which included Rikkert, Van Lohuizen and Schaad as members. Engberts (1973) maintains that, from the start, the committee was prejudiced against the Halle method as advocated by Amsterdam. Van Lohuizen and Schaad came from municipalities that had renounced further use of the method.[82] Van Lohuizen, moreover, was the inventor of an alternative. The report of his Rotterdam method was published by Hudig's NIVS (Van Lohuizen, 1922).

---

80 Among these municipalities were Haarlem, Nijmegen, Rotterdam and Zwolle (Van der Kaa, 1925). Other municipalities, like Rotterdam, The Hague and Schiedam, had worked out various variations of the method (NIVS, 1929). Van der Kaa criticism was published in the Dutch Journal for Housing and Townplanning, the voice of the Netherlands Institute of Housing and Town planning (NIVS). Van der Kaa had some reserves as to the general applicability but he did not want to renounce of its use, because the Halle method provided a simple way of getting insight in current housing need (Van der Kaa, 1925). Not only Rikkert but also Van Zanten wrote in defense of the method. Van Zanten was not without criticism of the method, but he thought that a great part of Van der Kaa's objections, namely that the yearly bookkeeping of 'entities in need of a dwelling' resulted in a cumulation of errors, could be prevented by periodical corrections with the data of housing censuses (De Gans, 1994a).

81 The debate reopened in the 1970s (Engberts, 1973).

82 At that time Schaadt was working at the Department of Building and Housing Supervision of Zwolle; Van Lohuizen at the Housing Department of Rotterdam.

Van Lohuizen's Rotterdam method did not differ in principle from the Halle method, only it was much more labourious and it asked for better and more elaborated statistics. Even Van Lohuizen admitted that his Rotterdam method could not substitute the Halle method in most small municipalities and in the big municipalities only in so far as the municipality can dispose of an exact information of household movements (Van Lohuizen, 1922). The Halle method and the method of Rotterdam have the assumption in common that each household needs a dwelling. In the Halle method changes of housing need are calculated from marital state transitions and migration. Each transition of marital state has a direct influence on the housing need. The Rotterdam-method was based on two years of experience with new dwelling statistics. Of each separate transition it was examined whether it had resulted in the disappearance of an existing household or in the formation of a new one (Van Lohuizen, 1922). It is another indication of the significance that was attributed to the housing issue and in the great efforts and investments of time and energy in the accurate estimation of the number of families (households) and the housing need in municipalities. Apparently Van Lohuizen's time consuming method never played a major part in housing need estimations. It is not recorded in Van Fulpen's thorough overview of eighty years of research of the expected growth of the housing stock (Van Fulpen, 1985). On the other hand it is referred to in Giedion's overview of the state of the art of international town planning at the end of the inter-war period. According to Giedion as far back as 1920, the city of Rotterdam used a carefully differentiated method for accurately determining not only the number of dwellings needed but the kinds of people who would occupy them and the sort of dwellings each would require (Giedion, 1940/1966).[83]

It cannot be denied that the committee was in a difficult position, given its composition. Rikkert, who was almost the incarnation of the Halle method, was one of its members. Hudig's post-war successor as secretary-director of NIVS, Van der Weijde, was presumably right in saying that the committee had only demonstrated a strictly critical mind; *'taking a different position was unthinkable'* (Van der Weijde, in: Engberts, 1973, p. 68-69).

---

[83] It is noteworthy that Van Lohuizen's paper was published in German. Giedion's overview demonstrates that international publications pay off in being inserted in intellectual histories of the development of a field of science.

The majority of the committee, opposed only by Rikkert, concluded that the accuracy of the Halle method could not be demonstrated with the census data of Amsterdam, as its advocates had said, because of differences of definition of households in need of a dwelling (NIVS, 1929). Another major criticism was the method's definition of entity-in-need-of-a-dwelling, which was taken quite broad. Brothers and/or sisters living together and other cohabiting combinations of non related persons were implied in the definition; in fact all non single person households were understood as family households. In everyday practice the range of household types is broader: There are also households of unmarried women, of unmarried or divorced men with or without servants, of cohabiting unmarried brothers and sisters, and so on. Deviations of observed trends are most considerable in unusual circumstances (crisis, epidemics, fast population growth, war), that is in times wherein accuracy is most necessary. Another criticism of the Committee was that household and 'entity in need of a dwelling' were taken as being synonym (NIVS, 1929). Not all households are in need of a dwelling or can afford a dwelling. The economic situation exerts an influence on housing need. For that reason housing need as calculated with the Halle method was redefined in *theoretical housing need* in Amsterdam in the mid-thirties (Nota van Toelichting, 1935). After the war this new definition was taken over by the housing expert Van Beusekom (in 1946; see Van Fulpen, 1985).

It took four years before the report of the committee was published (NIVS, 1929). Interest in the issue of the accuracy of the Halle method had subsided by then, as Van Zanten's review (1929) makes clear. He considered the fuss made by NIVS on the utility of the Halle method unintelligible and the activities of the committee a superfluous exercise. The report came too late in the day; the uproar had subsided. By then it was clear to everybody who had ever been involved with the method that it was based on a number of assumptions that did not completely coincide with reality. It was also known that errors made would compensate for each other to some degree. Advocates of the Halle method were also aware of its limitations, but saw no reason to stop using it because nothing better was available. Of course, verification counts were necessary from time to time so that the number of families (households) estimated by the Halle method could be compared with the observed number.

Van Zanten wondered why the method had been given the honour of *'such a momentous committee investigation'*. Somewhat sarcastically, he presumed that because the significance of housing was so great, calculation of the shortage of houses needed to be *'accurate, perfectly accurate indeed'*. He agreed with the committee's conclusion that the deviations were greatest in unusual circumstances of time, but he had his doubts on the relevancy and necessity of such a high degree of exactitude. The advocates of such accuracy could better take into consideration the difficulty of providing for a calculated shortage within one or two years. An accurate estimation of the total shortage, moreover, had little relevancy as long as it was not specified by the various tenancy categories (De Gans, 1994a, p. 22-23). The only valuable use of the committee's work, thought Van Zanten, was that it had investigated all the different methods for the determination of the housing shortage as had never been done before. Everybody having access to the unpublished annexes to the report was now capable of determining the value of each of the methods investigated (De Gans, 1994a).

6.4 | Innovations in the 1930s

Housing need forecasting methodology received two different innovative impulses in the next decade. They came from planners/demographers working in Rotterdam and Amsterdam respectively and resulted from the issues town planners had to deal with. Urban and regional planning had become well established activities in the 1930s. Town planners were convinced that a plan had to serve the needs of several future generations (Angenot, 1935). Taking into account the time it takes for a plan to become reality, it is clear that future social and economic development and the future growth of population were considered to be of paramount importance and dominant in the question of town planning. Recent forecasts of the population of the Netherlands carried out by the town planning offices of the three big cities had resulted in general agreement that the population of the Netherlands would reach a total of about 11 million inhabitants in 1970 and that the population would be relatively old by then. Town planners had to think about the consequences for the future size of the housing stock. Town planners like Angenot were aware that the observed relationship between population increase and extension of the housing stock was closely linked to the town and regional plans being prepared, and that the future production of new dwellings was not only a physical planning problem, but a housing and economic problem as well.

The housing stock increased much faster than the population. Apparently, as Angenot demonstrated, there was yet another factor at work instigating an extension of the housing stock: The decrease of the average family size (*gezinsverdunning*). From his own forecast of 1934, Angenot knew that, in the course of the process of transition towards a situation of low birth and mortality rates, the age structure of the population was changing. It would take about forty years to reach a stationary situation (in terms of stable population theory). In the meantime, the proportional increase of the middle and higher age groups would affect the economy and housing. For the time being new dwellings would have to be built on a relatively large scale, because of the increase in numbers of the family founding age groups (Angenot, 1935). Accurate forecasts of future housing need made it necessary to account for family size, age structure and marital state in the calculation procedure. Angenot found a mathematical equation that fitted developments over the period 1900-1930 so well that it took *'the shape of an historical law for that period'* (Angenot, 1935, p. 199):

$$H = m/2 + u + s - v$$

H the number of dwellings needed (the housing need);
m the number of married persons up to 72 years;
u the number of incomplete families (one parent families; multiple person households, et cetera) (equal to 25.3 per cent of persons from 25-72);
s the number of persons living separately;
v the number of inhabited vessels and caravans.

The values of the parameters of the equation came from observations of past developments. The 1934 population forecast of Rotterdam, discussed in the next chapter, formed the basis of Angenot's forecast of future housing need. He assumed constant age specific proportions of married persons at the level of the census of 1930, and constant values of the parameters of the equation during the forecasting period. Uncertainty with respect to the future number of births led him to use a minimum and a maximum population forecast. He was aware of the uncertainty with respect to the future behaviour of the parameters u, s, and v, but since their influence only affected about one fifth of the housing stock the consequences would be limited. From his calculations he could deduce the future course of average dwelling occupation figure (see *Figure 6.2*).

Vignette 11: Laurent Hubert Joseph Angenot (1901-1979)

When the Netherlands Central Bureau of Statistics took up the task of population forecasting and to publish about it, in the early 1950s, it was assisted by an advisory committee. With two exceptions the members of the committee were representatives of directly interested parties like the Central Planning Agency, the Ministry of Reconstruction and Housing and the National Agency for the National Plan. The exceptions were Angenot and Van Lohuizen, both engineers and graduates of Delft Technological University. The professional careers of Angenot and Van Lohuizen show similar patterns, both with professional careers that began in Rotterdam. They were advocates of the survey-before-plan doctrine in town planning and considered the population forecast as the founding stone of physical planning. They shared a rather positivistic attitude and were convinced of the necessity of quantitative foundations of town planning: it would help town planners to present their viewpoints in an exact way; similar to technicians and economists, who always have figures near at hand. Both were innovators of municipal forecasting methodology in the early 1930s.

Angenot graduated as a civil-engineer (hydraulic engineering) in 1926. After his study at Delft he obtained a bachelors degree of Dutch law at Leiden University, where he followed also courses of economy; he studied sociology at the Municipal University of Amsterdam. In 1930 he got a position at the newly established Division of Town Development at Rotterdam where he was put in charge of town planning research and matters of regional planning. Soon after he was commissioned to investigate what the consequences of the demographic trends of the Netherlands for the future population size of Rotterdam Harbour Area was likely to be. The results were published in 1934. His book is one of the most innovative works on forecasting methodology of the inter-war period.

In 1936 he was appointed chief of a division of, among others, townplanning and housing research. After the bombardment of Rotterdam of May 1940, wherein all documentation and the study-archives were destroyed, Angenot became Chief of the Division of Plan Research of the Advisory Bureau Town Plan of Rotterdam, a combined national and municipal bureau.

Like Van Lohuizen Angenot was an fervent advocate of the unity of the town planning. His unwillingness to disconnect research and design and his inclination of giving priority to demographic rather than to economic preliminary research, cost him his leading position in 1944. When the reconstruction plan of Rotterdam was made, Angenot was manoeuvred on a side track. A plan that resulted from extrapolation of pre-war (demographic) tendencies, which was preferred by Angenot, was thought to be insufficient as a guideline for the future. The future position of reconstructed Rotterdam had to be the result of the visions of planners, managers and politicians.

Angenot was greatly involved in demography. His demographic interest dates from the time of his study in Delft. In 1920 he attended Prof. J.A. Veraart's lectures on the population problem, an issue with a high degree of actuality at that time. Veraart was a fierce anti-neo-Malthusian who denied the very existence of a population problem, because the

Prof. L.H.J. Angenot

world was big enough to feed a growing population. Notwithstanding his criticisms of Veraart's way of reasoning, the interest of Angenot in population questions was kindled. His demographic publications deal with relations of population growth and urbanisation in particular. His 1934 population forecast brought him in the ranks of the leading population forecasters of the nation. In 1950 he became the first chairman of the Netherlands Demographic Society.

As a town planner he had international renown. As a demographer he belonged to the few Dutch demographers with an international reputation in the Fifties and Sixties (Source: De Gans, 1997)

Basically, Angenot's approach was a simplification of the Halle method. He transformed it from a potentially dynamic forecasting method into a more static method that could be used more easily, but with the advantage that the parameters could be made dynamic easily were it necessary to do so. This could be done if necessary by the exogenous attribution of a specific value per forecasting interval. As Rikkert had done 15 years before, Angenot was able to demonstrate that the average occupation figure depended on population size and composition. Angenot did not explicitly arrive at this conclusion in 1935. But implicitly his approach confirmed that the average dwelling occupation figure was not suitable as an independent forecasting instrument.

Almost simultaneously, and for the first time in the Netherlands, Van Lohuizen and Delfgaauw, who were Angenot's counterparts in the town planning team working for the General Extension Plan of Amsterdam, developed a method known after the war as the age specific headship-rate method. The method was developed for the calculation of the future number of new dwellings for which the extension plan had to provide. The number was thought to be determined by the increase in the number of families, the replacement of uninhabitable and unsanitary dwellings by new ones, and by the substitution of dwellings demolished or having lost their housing function, through the extension and depopulation of city centres (Nota van Toelichting, 1935). An analysis of the observed trends of the average dwelling occupation figure for the whole town and various neighbourhoods produced a conclusion similar to Angenot's, namely that the number of dwellings had increased because of population growth and a decrease in the average dwelling occupation figure as a consequence of changes in the composition of the population. Once the age-specific headship rates were known and a population forecast by size and age was available, the future number of households could be calculated as a basis for the calculation of future housing need.

The forecasters of Amsterdam started from the so-called 'Halle families' (the numbers of married, divorced and widowed women). From the population census of 1920 the proportions of women of the distinguished age categories were known. These proportions were assumed constant over de forecasting period. The population forecast of 1932, of which more in the next section, provided the future numbers of women of each age group at the end of each ten year interval. With these data the theoretical housing need, total and by age group, could be calculated. This figure had to be corrected by adding the housing need of other persons, men in particular. From the dwelling census of 1925 it was known that the housing need of these persons amounted to about five per cent of the theoretical housing need of women (the theoretical housing need was equal to the housing need of the Halle families) (Nota van Toelichting, 1935).

Figure 6.2. Average dwelling occupation rate of the Netherlands, 1899-1970, observed and calculated

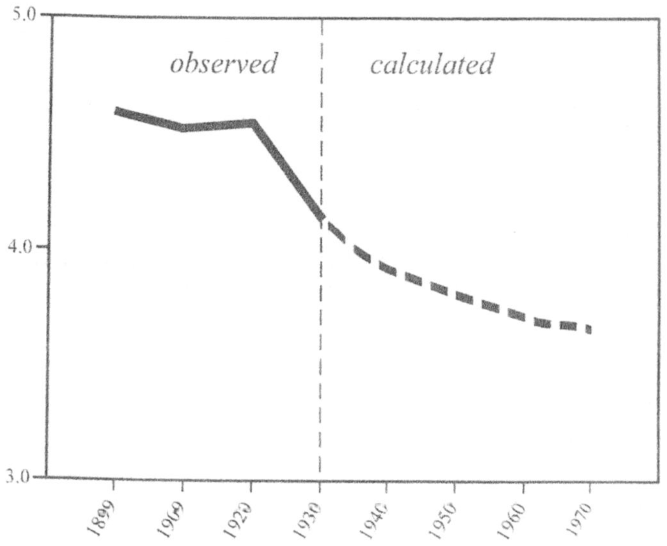

Source: Angenot, 1934

Angenot's method did not necessarily demand a population forecast by age; it took account of changes in the marital state composition in the first place. On the other hand, in the Amsterdam method the influence of changes in the age composition on the housing need were clearly accounted for and explained.

6.5 | The State-of-the-Art Prior to World War II

As a result of developments in the methodology of calculating the (future) number of households and future housing need, Dutch forecasters could avail themselves of an arsenal of forecasting methods at the end of the inter-war period. They had insight into the contribution of the new population forecasting methodology in understanding changes of the future housing need and the relative utility of the average dwelling occupation figure in forecasting future housing need. The standard average dwelling occupation approach was never completely abandoned. The dangers implied in its application had been clearly exposed by Rikkert (in 1919) and Angenot (in 1935); Rikkert by explaining the influence of changes of the processes of family formation, Angenot by emphasising the influence of age structure and marital state composition of the population.

The second method was the Halle method. Although mainly used for estimating current housing need, it provided the analytical demographic basis of a process oriented (that is, a dynamic) approach to family (and household) formation.

Third, there was Angenot's marital state composition approach which in principle made it possible to account for changes in the marital state composition of the population and the age structure of the population.

Lastly, there was the Amsterdam headship rate method which accounted directly for changes in the age structure and, on the other hand, asked for a demographic age specific population forecast in its most explicit form.

The discussion of the merging of demographic and housing need forecasting in the 1930s, however, anticipates the discussion of what happened in the field of urban population forecasting in the preceding decades. This point is taken up in Chapter 7.

7. The Search for Practical Applications in Dutch Urban and Regional Forecasting

'We do not search after a certainty for the future that cannot be given, but after the most likely hypothesis. We cannot obtain more; we should not be satisfied with less. The more profound our investigations, the more precise our assumptions for the future, the better our plans.'

L.H.J. Angenot in 1934[84]

7.1 | Influences from Abroad on Urban and Regional Forecasting

In contrast with the context of national forecasting, which to a high degree remained the domain of scientific discourse of the issue of future population, urban and regional forecasting were not a matter to be dealt with without engagement. Here it is not scientific interest in the population problem which was the stimulus to methodological innovation, but the context of urban policy and decision making. Policy making required more precise information with respect to investments to be made and demands to be met. Forecasters were not in a position to discuss the appropriateness of speculative demographic forecasting in the absence of a law of population growth and as members of

---

84 In his course of town planning instruction at Delft, December 17, 1934 (Arch. Angenot, inv. nr. 314, NAi).

urban town planning teams they could not – as Van Zanten thought he could afford to abstain from making speculations on the future course of urban population. This accounts for the significance attributed to population forecasting in town planning. The urban housing context, as discussed in the preceding chapter, accounts for the efforts vested in the accurate estimation of the current housing need. Urban and regional planners, however, were particularly interested in long-term population forecasts and the consequences of future population development for housing and other needs.

The great transformation of population forecasting in the Netherlands took place in the 1920s when Wiebols and 't Hooft laid the foundations of the demographic approach in the study of the future course, size and structure of the national population. However, the further advancement of the methodological innovation of demographic forecasting came from town planning in the 1930s. It became clear, as for instance Bakker Schut indicated his 1933 forecast of the Netherlands and The Hague, that many domains of municipal policy making could profit from population forecasting.

The innovation of inter-war municipal forecasting was the work of professionals involved in urban planning. In comparison with national population forecasting, municipal forecasting was not the result of the action of private persons with primarily a scientific interest in the population problem and the calculation of future population. In municipal planning, for instance, general extension plans had to provide for new housing opportunities, for the basis of investments in social housing, infrastructure and other fields of municipal and regional policy.

Large traffic surveys, population forecasts, studies of housing need and the need of recreation areas and industrial parks were conducted for the first time in the Netherlands in the late 1920s. Such studies had been put under discussion by the German town planner Baumeister about half a century earlier, in 1876 (Angenot, 1973). German town planners were decades ahead of their Dutch counterparts. That they influenced Dutch town planning is no wonder, for Germany experienced explosive urban development well before the Netherlands, where the Industrial Revolution, explosive population growth and urbanisation lagged a few decades behind. German manuals, acts and practice examples left indelible marks on Dutch thinking on the extension of towns in the first decades of the 20th century (Angenot, 1973; Van der Valk, 1989). Urban and regional population forecasting as part of preliminary town planning research were influenced by the German authors on town planning, Baumeister and Stübben in particular.

Baumeister was convinced that municipal population growth was determined by the geometrical law. However, his approach to the law was pragmatic. His conviction was founded on an empirical study of the increase of population of the most important towns of Germany in the period 1843-1871. The generality of the law of population growth had not, according to Baumeister, to be taken too seriously, for in some towns population growth was slower, in others faster, and sometimes population growth was interrupted. Forecasting population growth is, therefore, a problematic affair. The purpose of town extensions is to provide for the future. It is advisable to develop an approximate law of population growth from an analyses of statistical data from the past. If necessary, this approximation can be modified. That the law of population growth will remain constant cannot be taken for granted. Once the growth rate is known, however, the doubling time of the population can be readily calculated. When a town extension is envisaged, the first activity to be undertaken is a statistical investigation of population growth, births and deaths, immigration and emigration according to age, profession, income and with respect to provisions. The data can support the drawing of conclusions on the future development of the town and on the needs of a town extension. That the expected development is actually realised is by no means certain, because the expectation is based on statistics (Baumeister, 1876).

Fourteen years after Baumeister, Stübben propagated similar ideas on future population growth in his town planning manual *Der Städtebau*. He pointed at population increase as the main reason for the extension of existing towns and the building of new ones and advocated prudent use of the method of geometrical population growth, particularly for the calculation of the doubling time of populations. Prudence was advised, because housing demand lags behind population growth in periods of economic recession, while proportionally more dwellings are needed in periods of increasing prosperity. Stübben was a propagator of the use of standard figures in town planning, including standard figures related to future population growth. If, for instance, the population of a town increases at a rate of four per cent a year, then 2.5 times the present surface of the built-up area has to be reserved when a plan for a future of 25 years is envisaged; 1.5 times the present expanse in the case of a growth rate of two per cent a year (Stübben, 1890).

The preliminary reports of the annual meeting of the Society of Political Economy and Statistics on *'the line of action to be recommended to municipalities in order to further appropriate extensions of the built-up areas of these municipalities'* provide evidence of the German influence on town planners in the Netherlands: The reports abound with citations from German publications (VSS, 1908). The courses on town planning law given by Valckenier Kips were organised largely on the lines of German examples (Van der Valk, 1990).[85] In fact, Valckenier Kips became the highly influential intermediary between the German innovators and his own town planning students, including such civil engineer/town planners of the first generation as K.T. van Lohuizen and M.J.W. Roegholt (Van der Valk, 1990; Roegholt, 1925). In his publications and courses he stressed that, before starting a town extension plan procedure, a town planner had to know the rate of growth of the total population and the various population categories. He taught his students the sense of using standard figures, for instance regarding the number of rooms per family, and the width of roads according to their function (VSS, 1908; Stolzenburg, 1984; Faludi and Van der Valk, 1994).

Valckenier Kips' view of the prediction of future municipal population was similar to that of Baumeister and Stübben. It was based on the concept of geometric growth and the doubling time of a population. As to the future composition of the population, Kips was especially interested in the future social economic structure of the population. Valckenier Kips made quite an impression at the Annual General Meeting of VSS in 1908 for which he had written a preliminary report. This becomes clear, for instance, from a summary of the proceedings of the meeting written by Faber: *'Mr Kips traces the technical exigences of an appropriate town planning, whereby the concept of 'the near future' as used in the Housing Act is interpreted as 'for a doubled population or for an area as big as the existing built-up area'. This is arbitrary, but from a practical point of view there is much to be said for it; the Act itself is quite vague at this point. The first question that arises when a plan is designed is: For what kind of people must be built; how will the new population be distributed over the various social categories; will space be*

85 J.H. Valckenier Kips was a professor of constitutional and administrative law at the Delft University of Technology (from 1909-1932). He was appointed in 1909, presumably because of his many publications on housing and town planning matters. Valckenier Kips was influenced by Stübben and Sitte; he was the sole professor of town planning in the Netherlands at that time (Rückert, 1914).

needed for worker families and for industrial enterprises mainly or are districts for more substantial citizens, and villa districts expected also? Miscalculations at this point will be revenged seriously' (Faber, 1908, p. 90). As late as 1925, a former student of Valckenier Kips referred to his contribution, stating that preparation of an extension plan asked for information on the future growth by population category of the municipal population as Valckenier Kips, who had been the first in the Netherlands to put this very important demand into words, had advised. Not only must the coefficient of increase of the total population be known, but also the coefficients of growth of the different classes of the population.[86]

7.2 | 'Survey Before Plan'

In the middle of the third decade, British influences started to manifest themselves in the Netherlands through the *survey before plan* approach. These reinforced the effect of the earlier German influences. It was the conference of the *International Garden Cities and Town Planning Association* of 1924 which familiarised the Dutch with the concept. The conference was part of a carefully orchestrated campaign to promote the issue of regional planning and to settle the issue of garden cities around Amsterdam by Hudig, the director of the National Institute of Housing and Town Planning (NIVS). The issue of garden cities came to nothing, but the conference familiarised the Dutch with Geddes' doctrine of *survey before plan* propagated by Abercrombie and Unwin, and it helped the idea of regional planning among policy makers to mature (Faludi and Van der Valk, 1994).[87] Compared with Baumeister and Stübben, the idea of a survey was not new, but as a concept *survey before plan*

---

86 Roegholt (1925) in his PhD thesis on the metropolitan region. The reference is not quite surprising: Valckenier Kips was Roegholt's supervisor.

87 The idea came from Ebenezer Howard's Garden City Movement. Originally garden cities were meant to lighten the population burden of big cities. Later it evolved into a concept of suburbs and new town extensions of central cities, with aspects of regional and metropolitan planning coming to the fore. In 1913 the *International Garden Cities and Town Planning Association* was established. It changed its name in 1924 in *International Federation for Housing and Town Planning and Garden Cities* and in 1929 in *International Federation for Housing and Town Planning* (Bosma, 1993). Hudig was the thriving force of NIVS' joining of the International Garden Cities and Town Planning Federation. NIVS succeeded in manifesting itself both national and international by undertaking the organisation of the *International Town Planning Conference* in Amsterdam in 1924.

captured the imagination of planners. It became a powerful planning principle of first half of this century (Faludi and Van der Valk, 1994).

The Scottish biologist Patrick Geddes (1854-1932) was, if not the inventor, then at least the propagator of the concept of *survey before plan* (Geddes, 1915/1949).[88] He introduced it at the Town Planning Conference of London of 1910 (Angenot, 1973). Geddes could build on the recent British tradition of civic surveys in London (by Booth in 1889-1903) and York (by Rowntree in 1901) that described the living conditions of great parts of the population. His most important contribution to town planning was the introduction of the regional survey method. Geddes was particularly interested in the analysis of the transition of the region as a natural area into an intellectual area, the cultural evolution of the region and the part played by mankind and technology. He investigated the stages of development of the towns in the region and the significance of the urban heritage in view of a *'useful past'* (family life, the relationship with nature and the integration of the individual in the local and regional community). Starting from the present, he searched for the trend lines of past developments in terms of laws of development (the 'useful past') that can be extrapolated into the future. Given the results of the social survey, Geddes thought it possible to make accurate predictions of the future and to arrive at city development. To that order the investigator had to demonstrate insight into the actual evolution of towns and a sharp eye for the germs that would play a part in the future (Bosma, 1993, p. 109). "*Past origins, present facts, future developments are thus considered, and for the people's life, as well as for their homes*" (Geddes, 1915/1949, p. 92). His approach was a comprehensive one, for the "*whole situation and life of a community in past and present*" is surveyed. The future was to a large part decided by the plan: It prepared "*for the planning scheme which is to forecast, indeed largely decide, its material future*" (Geddes, 1915/1949, p. 125).

His ideas were positively received in the United States of America, particularly in New York. In 1922, the Trustees of the Russell Sage Foundation collected a considerable amount of money for a survey that would precede the making of a Regional Plan for New York. Thomas Adams, a friend of Geddes and one of Ebenezer Howard's collaborators at the establishment of the Garden City of Letchworth, became the director of this survey (Angenot, 1973).

---

88 Geddes was influenced by German examples (Geddes, 1915/1949; Faludi and Van der Valk, 1994, p. 48).

Adams, was one of the preliminary advisors of the International Town Planning Conference in Amsterdam in 1924. Geddes' influence can be traced in several Dutch overviews of topics to be investigated in surveys as part of preliminary town planning research (De Casseres, 1926; Leidraad, 1927). It can easily be recognised in the rather fixed schemes of Dutch sociographic studies of rural municipalities used during and after the Second World War (Stolzenburg, 1984). As far as population is concerned, Geddes was interested in its movement and in resulting requirements such as housing need, schools, occupations, health, density, distribution of well-being, education and cultural agencies (Geddes, 1915/1949).

7.3 | The Housing Act of 1901

The Housing Act of 1901 provided the legal basis for town planning, in general extension plans in particular. From 1901 onwards, town planning interventions were only possible in the form of extension plans (Rossen, 1980). This is not to say that extension plan making was a totally new activity in the Netherlands. From the beginning of the 19th century one comes across the concept of 'regular town extension' (Van der Valk, 1989). Van der Valk asserts that the metaphorical use of the adjective 'regular' is evidence that an extension plan was seen as a picture of the future spatial situation. The earliest Dutch publication in which the principles of regular town extension were formulated, the text of a lecture by the state architect W.N. Rose, dates from 1862 (Van der Valk, 1989; Wagenaar, 1992).

The novelty of the Housing Act was that forecasts of future population, housing and extension plans were linked in a direct way. Originally, the plans were designed to prevent such disorderly situations as had arisen at the end of the 19th century because of the explosive population growth. Gradually, an evolution towards positive, constructive action took place. Town plans started to take on the character of a programme intended to direct the development of a town *'with a firm hand along rational lines'*.[89]

With respect to the development of urban population and household forecasting, the Housing Act is significant in two different ways. First, municipalities

---

89 Angenot, in a course of town planning instruction at Delft, 1934 (Arch. Angenot, inv. nr. 314, NAi).

had to have a clear insight into the existing housing need and the course the housing shortage was taking. They had to make estimates of the (future) demand of houses by households. This resulted in a growing interest in the development of good methods of estimation. Second, German publications and examples had convinced the first generation of Dutch town planners of the necessity for insight in the future course of the municipal population to precede the making of an extension plan (Angenot, 1973; Van der Valk, 1990).

The coming into operation of the Housing Act did not, however, mean that municipalities were eager to make extension plans. There were quite a few complaints of municipal obstruction. Many municipalities *'apparently were too much behind the times to be able to bring into practice the high intentions of the legislator'* (Van der Flier *et al.*, 1952, p. 49). In 1911 the Chief Inspector of Public Health, H.J. Faber, lamented that not much proof of enthusiasm among municipalities could be detected. Various municipalities immediately tried to get exemption, and others did not even try to make an effort, but just kept silent, presumably thinking *'.. that any obligation for which there is no sanction is not an obligation at all'* (Faber, 1911, p. 286).[90]

Of course, there were exceptions. In some cases experts were commissioned to make a plan. In other cases, including some of the most important municipalities, plans were made and even agreed upon, but in many instances practice differed from appearances. The attitude of municipal authorities towards the new obligation was often rather cynical (Faber, 1911). Faber gave an interesting glimpse behind the scenes: *'As a rule the director of Public Works was asked to fix the job in his leisure-hours, and in doing so the significance of the plan was degraded, which naturally often showed out. In case provincial authorities objected to and disapproved of the plan in the end, it would of course be a pity because of the amount of work it had taken, but at the other hand it would not be too bad; one had done one's best and the plan could as well be considered to be approved of by the provincial States Deputy, because they were not in the situation to do something about it'* (Faber, same place).

90 J.H. Faber (1864-1921) was one of the four Chief Inspectors of Public Health who superintended the performance of the Housing Act. State superintendency of the Housing Act fell under the Health Act of 1901. Inspectors of housing were appointed after the alteration of the Housing Act. Faber is referred to as Health Inspector and as Housing Inspector (Van Beusekom, 1952; Van Sandick, 1921, p. 519).

It took four more years for a serious attempt to be made to integrate social-scientific knowledge in a general extension plan. The attempt came from Rückert who worked for the municipality of Tilburg (1917). For a long period, his plan was a *rara avis* among general extension plans. While the making of extension plans started to become a normal town planning activity after World War I, the thorough socio-economic or demographic foundation was mainly limited to the domain of theorising on planning, but was rarely put into practice (Stolzenburg, 1984; Van Kesteren, 1984).

7.4 | Towards a Three-Tier Planning System

The extension plans of the municipalities of Velsen (of 1922), a municipality of about 29,000 inhabitants in that year, and of Nieuwer-Amstel in 1923 and 1924, were designed for absurd population numbers; 250,000 and 120,000 inhabitants respectively. The plans were mere street plans, or capacity plans, with only a shadow of the argumentation based on the kind of research advocated by Valckenier Kips (Van Kesteren, 1984).

The effect of these plans was a call for regional and metropolitan planning. The increase in municipal extension plan activities created problems of their own, since municipal autonomy started to pose obstacles to rational planning at the regional level. In many instances there was a lack of willingness among neighbouring municipalities to co-operate in adjacent extension plans. A regional approach to town planning was required. At first, regional planning was seen from a negative point of view. It had to serve as an instrument to prevent 'bad neighbour' problems from happening, or to solve them, including parasitic behaviour by suburbs of big cities. In the latter case, regional planning was seen as an alternative to annexation. Gradually, however, a more constructive approach prevailed. A regional plan came to be considered as a comprehensive form of planning for an area instead of an instrument to solve problems (Faludi and Van der Valk, 1994).
One of the thriving forces behind the advancement of modern regional planning in the Netherlands was Hudig.[91] In the mid 1920s, after the Interna-

---

91 Dirk Hudig (1872-1934) started his impressive career in social housing as a member of the Committee for Housing in the 'New Town' of Amsterdam, which was commissioned to investigate the housing situation of Amsterdam and to make proposals for the improvement of the housing situation. Hudig was one of the founders of the national

tional Town Planning Conference of 1924 in Amsterdam and the publication of Roegholt's dissertation on the metropolitan region in 1925, the debate on the feasibility of regional planning was in full swing.

The Velsen plan had another side effect with implications for population forecasting. The controversies raised by the plan resulted in heterogeneous recommendations to the States Deputies of the province of North Holland, who had to give their assent to municipal extension plans. The States Deputies decided to put an end to the controversy. The call for regional planning, for the safeguarding of natural beauty and the provincial need of unanimous advice resulted in the establishment of a standing advisory committee for the evaluation of extension plans, first in the province of North Holland and later in other provinces (Van Kesteren, 1994).

This Standing Committee for Extension Plans in North-Holland was established in 1925 and started its activities in 1926. Hudig, the secretary of NIVS, became its deputy chairman. One of the tasks of the Standing Committee was to instruct municipalities on the design, organisation and implementation of extension plans. The publication of its directives for the included directives for forecasts. The set of guidelines became the first manual of physical planning, while the forecasting directives were the first attempt to standardise local forecasting methodology (Leidraad, 1927).

Provincial planning became one of the three building blocks of the nation's three-tier planning system (national, provincial and municipal), consisting of a hierarchy of municipal, regional and national plans. The foundations of the system were laid in the inter-war period. In 1939 Kloos demonstrated the necessity of a national plan as the finishing touch of the Dutch system of physical planning in his PhD on the national survey, fourteen years after Roegholt's book on the regional (metropolitan) plan (Kloos, 1939).

The advocates of three-tier planning system exploited their opportunities during the German occupation of the Netherlands in having a hierarchical

---

federation of housing corporations (the National Housing Council; in 1913). He was one of the main initiators of the foundation of the National Institute of Housing and Town Planning (NIVS) in 1918 (Van der Schaar, 1980; De Ruijter, 1987). From 1923-1934 he was its secretary-director. It's Town Planning Council (1923) dominated future discussions on regional and national planning.

structure imposed that was at odds with the existing system of municipal autonomy (Faludi and Van der Valk, 1994). A Government Service for the National Plan was set up in 1941. It was the forerunner to the National Physical Planning Agency. It came into full swing after the Second World War.[92]

In 1943, the Provincial Standing Committees were abolished by the German occupiers and the Provincial Physical Planning Departments (PPD's) were established in all provinces instead (Van Kesteren, 1984). By the end of the period, a Division of Town Development or Town Planning was in operation in many a large town. This marked the beginning of a new era in planning. As Faludi and Van der Valk (1994, p. 74) put it: "*Previously planning had been a mere adjunct of housing. Now planning considerations were to come first, and housing (and other matters) only afterwards.*"

7.5 | Professionals in Town Planning

It took long for the ideas of Baumeister and Stübben, as taken up by Valckenier Kips (in 1908), to be put into practice. It is a matter of discussion why this was the case. The struggle among town planners in the first decades of the century that resulted in a shift of 'school dominance' in town planning may afford an explanation. Rückert, an active participant described the struggle as between the 'utilitarian' school (mainly military and civil engineers) and the 'city beautiful' school (the "*schoonheidsschool*", mainly architects). In the end the architects won (Rückert, 1913; 1914). While the civil engineers followed Baumeister and Stübben, the architects took their lead from Camillo Sitte, who stressed the importance of aesthetics in town planning and fought both the primacy of engineers and their obsession with the 'regular extension' of built-up areas. Sitte, for example, accused the civil engineering approach

---

92 Frits Bakker Schut, a civil engineer and former Housing Inspector, was the first director. He was taken by the idea of a comprehensive National Plan as the top of a national planning hierarchy running from a national plan, via regional plans to local plans. The basis of Bakker Schut's thinking was the distinction between planning and spatial ordering now commonplace in the Netherlands. 'Spatial ordering' refers to the concrete measures taken by authorities with respect to the development of the national, regional or local region; 'planning' is the co-ordination of these actions (Faludi and Van der Valk, 1994, p. 77).

of furthering the ugliness of towns. The expression 'engineer plan' became a pejorative term applied to an ugly town plan (Van der Valk, 1990).

In the practice of town planning, however, lines were not so clear cut as this typology might suggest. H.P. Berlage for instance, one of the most influential architects and town planners in the Netherlands in the first decades of the 20th century, was a fervent advocate of the architect approach. He was influenced both by Stübben in the early years of his professional career, and by Sitte whom he followed rigidly first and more freely later (Van Rossem, 1988). Neither Berlage nor Sitte denied the importance of preparatory research for a town plan, including the assessment of population growth, traffic development and the demand for construction (buildings). But that part of the job was in their view the task of the engineer (De Ruijter, 1987).

It is not surprising that Rückert held a different opinion from that of the architects. In his view, an extension plan could be made by a team of three specialists; an economist/sociologist, an engineer and an architect. He was not, however, in favour of such an approach; a plan should preferably be made by one person. An engineer trained in aesthetics was the most likely candidate for the job, and not an architect since architects were not so experienced and familiar with large infrastructural constructions like the layout of harbours, canals and tramlines, nor with roads, bridges or sewerage systems (Rückert, 1913; 1914).

The shift towards a predominance of the architectural approach to town planning may have had a retarding effect on the appreciation, application and innovation of urban population forecasting methodology. Where architect/town planners were responsible, not too much interest in the social-scientific foundations of a plan could be expected. Until the 1930s, the only general extension plan in which a serious attempt to a socio-economic foundation can be traced is Rückert's plan of Tilburg of 1917.

A side effect of the increasing acceptance of preliminary research was an increase in the number of professions involved in town planning and planning research. Architect-town planners, now trained at Delft Technical University as modern town planners, took over the design part which is considered as the most important component of town planning. In 1917 Rückert, ambitious to make a general extension plan, could undertake it alone. By the end of the

1920s, his acknowledgement that a team consisting of an engineer, an architectand an economist/sociologist could also do the work had gained credence. Geo-graphers engaged in survey research were admitted to the Town Planning Council of NIVS at an early stage. The new town planners, responsible for design, and engineer-town planners, who elaborated the engineering aspects of a plan, no longer had the time for conducting the necessary surveying work (Van der Heiden, 1990). The three large cities of the Netherlands were the first to employ town planners primarily interested in preliminary town planning research: The engineer/town planners Van Lohuizen in Amsterdam and Angenot in Rotterdam, while Kloos, an architect by training, was contracted in The Hague.

The growing number of extension plans, the increase in demand for surveys for town plans and a gradual increase in regional plans resulted in an influx of social geographers, sociographers and economists in preliminary planning research in the 1930s. Professor L. van Vuuren of the Institute of Geography in the University of Utrecht, trained his social geography students in preliminary regional research. They were trained to make so called 'prosperity investigations'. These investigations were often part of regional plans, and often commissioned by municipalities and Chambers of Commerce. Similar activities were undertaken by sociographers of the Municipal University of Amsterdam under the supervision of S.R. Steinmetz and his successor H.N. ter Veen, and by economic geographers of the Economics University of Rotterdam under the supervision of W.E. Boerman (Stolzenburg, 1984; Bosma, 1993, pp. 189-193). By the end of the 1930s, they had succeeded in organising national social-scientific research through establishing research institutes such as the Foundation for Population Investigation in the Reclaimed Zuyder Sea Polders (1936) and the Institute for Social Investigation of the Population of The Netherlands (ISONEVO, 1940). ISONEVO developed into a research centre that contracted surveying work as part of preliminary town planning research commissioned by municipalities.

The influx of the new professionals in planning and social-scientific research institutes was viewed initially by the town planning establishment with mixed feelings. ISONEVO's competence as a research centre, particularly in economics, was questioned by the Standing Committee for Extension Plans and Regional Plans in North Holland and the recently established National Planning Office. In this connection it is interesting to note that the Standing Committee,

Vignette 12: J.H.E. Rückert B.Sc. (1879-1918)

Rückert was not a man to be wary of taking a stand. He considered his training as a military engineer at the Royal Military Academy of Breda to be on a par with the training of a civil engineer at Delft Technical University. This point of view was not always shared by his fellow civil engineers. In Rückert's view a general extension plan of a municipality had to be made by an engineer, either military or civil, and not by an architect. According to Rückert, architects, who were gaining ground in town planning, tended to focus unduly on the design aspects of a plan and knew little of the more technical aspects. In the struggle for dominance by various 'schools' in Dutch town planning in the first decades of the 20th century, Rückert was an advocate of the engineering approach (the utilitarian school), in opposition to the architectural approach (the aesthetic school), although he acknowledged that engineers should be qualified in aesthetics. In the period when Rückert was trying to stake out the town planning field of activity for engineers, he was too late; architects had already appropriated a substantial part. Moreover, the professional field had become more diverse. In 1900, a Department of Architecture was established at Delft University of Technology. From then on a new category of engineers was trained: Architectural engineers (the new town planners).

On graduating from Breda as an engineer, Rückert spent some years in the army. Instead of a lifetime career as a military engineer he decided to follow a different course in civilian life. From 1907 until 1912 he worked as an engineer of municipal works at Leiden. On 1 January, 1913 he started work as the first director in Tilburg's Department of Public Works which had been established through the reorganisation of the Department of Municipal Works.

Following the Housing Act of 1901, the province of North Brabant urged the municipality of Tilburg to make a general extension plan. For a long time the municipality persisted in refusing to make such a plan; it was considered to be unfeasible, because of the cobweb extensions that had taken place in the past. An attempt by the Director of Municipal Works in 1910 had been unsuccessful; the municipality considered it unsatisfactory. It was decided to hire an expert who could do better. Rückert was that expert. Within three months of his appointment he was commissioned to undertake a general extension plan. His plan was presented in 1917 and approved by the municipal council in April 1918.

At that moment Rückert had already left Tilburg to take up the position of Director of Public Works 's Hertogenbosch. He died soon afterwards, in 1918.

Rückert was an innovator in local population forecasting methodology, not through the novelty of the methodology of his forecast, nor because his forecast for 1969 proved to be so good, nor because he advocated the integration of social-scientific knowledge in town planning, but because he was the first to put such knowledge into practice. (Source: De Gans, 1997).

J.H.E. Rückert B.Sc.

although in favour of studies of the future (e.g. future housing demand), was of the opinion that such studies could be readily undertaken by the town planner; they were a normal part of the town planning task.[93]

Ambivalence can also be discerned in the views of some of the most outspoken advocates of preliminary town planning research, Angenot and Van Lohuizen, both Delft educated town planning engineers. Regional planning activities, said Angenot, were initiated on a large scale at the end of World War II. It was not possible to find town planning engineers interested in preliminary research in sufficiently large numbers. People with a technical training thought this aspect of planning quite marginal. Good economists were scarce; they could easily find interesting jobs in other sectors of society. In contrast, for social geographers and sociographers, planning research was a welcome outlet,

93 Letter Standing Committee of 29 September 1941 to the Governor of the Province of North Holland (R.A.N.H., Arch. Vaste Commissie, Doos 0052, dossier 1, 1941, Den Helder. Also the minutes of the meeting of the Standing Committee of 21 July 1941 (same place).

almost the only alternative to teaching. This source of researchers made it possible to fill the vacancies in town planning research. In the beginning, because of their theory directed training, several of them had difficulty in accommodating to the target-orientation of physical planning, especially when they had to work in an office as solitary investigators and were asked practical questions that had to be answered in a hurry (Angenot, 1973). Apparently geographers and sociographers were initially seen as second best professionals in the (preliminary) planning research job, ranking third after engineer/town planners and economists.

Angenot's view was shared by Van Lohuizen, who taught preliminary town planning research in Delft after the Second World War. Van Lohuizen thought social geographers lacked application-mindedness: One could not be satisfied with the analysis and description of surveys; designing a plan required looking ahead into the future. Sadly, town planners were not eager to participate in Van Lohuizen's courses on town planning research; his student audience consisted mainly of geographers (Van der Valk, 1990).

7.6 | Population Forecasting Prior to the 1930s

Already, at an early stage of the development of (town) planning, the significance of population forecasting for town planning purposes was hardly contested, as a result of the influence of Valckenier Kips and others. But the uncontested position of population forecasting did not imply that forecasters were initially eager to apply the achievements of Wiebols and 't Hooft in demographic forecasting the moment they were available.

The course of events in municipal population forecasting of the 1920s demonstrates that population forecasting for planning and decision making purposes has its own inner logic, which is not necessarily that of scientific interest.

The first to apply Valckenier Kips' ideas in a general extension plan was the military engineer J.H.E. Rückert, a man with a mind open to new developments in his field. Rückert consulted experts in specific subjects, in particular Valckenier Kips, author of the foreword to the report in which Rückert's Tilburg plan was elucidated, and relevant international literature. One of his first actions when appointed Director of the Department of Public Works of Tilburg (in 1913) and starting preparations for the general extension plan of

Tilburg was to update his knowledge of town planning. He participated in a course of instruction on town planning organised at the Building Exhibition of Leipzig by the Seminary of Town Planning at the Technical University of Dresden. There he became acquainted with the work of Baumeister and Stübben (Rückert, 1913, 1917; Stolzenburg, 1984).

His forecast of the total population of Tilburg was traditional. In the wake of Baumeister, Stübben and Valckenier Kips, Rückert applied the geometric growth approach to population and, like most of his contemporaries, he was mainly interested in the doubling time of Tilburg's population. The use of the geometrical growth approach conformed with the newest insights in town planning, a logical choice once he had decided to start from a 'plan capacity' approach. He subscribed to current international theory that extension plans should provide for a number of inhabitants at least twice the actual population. Its novelty rests in the prudence of its application: Time and again, Rückert looked for checks for his calculations and feasible arguments. He was inventive, although not an innovator in forecasting methodology, and was exemplary in his search for a solid foundation for his assumptions within a wider socio-economic context.

Rückert did not believe in a law of population growth. He was convinced that the deduction of future developments from past ones did not provide a solid (safe) standard; the future may bring rates very different from those calculated (Rückert, 1917).

His analysis and future expectations of the demographic development of Tilburg can stand comparison with those of Rooy of the Netherlands (1920, 1921) discussed in Chapter 6. Although his forecast was based on orthodox geometrical growth theory, it was demographic in its assumption making and comparable in quality to that of Methorst (1922). In the next 15 years the methodology and sense of self-criticism was not improved in any other known municipal forecast of the Netherlands.

It is interesting to note how reticent Rückert was with respect to the analysis of the fall of the birth rate. He circumvented the problem by restricting himself to the future course of the growth rate. The idea that fertility and procreative behaviour could be an object of social-scientific research was not self-evident until far into the 1960s. The prudent prefaces to a report of an investigation of future family size among Dutch women who had given notice of marriage in the early 1950s (Diels, 1951, 1953) and to the proceedings of a conference on the Population Problem held in 1966 are other demonstrations of this attitude (KNAW, 1966). There was strong opposition from religious and confessional political parties in the Netherlands against making fertility the object of scientific research. Rückert was a child of his time. His reticence can be explained either from his own religious background, or from the fact that he was working in a municipality where Roman Catholicism dominated public life. Bakker Schut's forecast of 1933 displayed a similar reticence.[94]

Apart from the novelty of his systematic check of the credibility of calculations and assumptions and the demographic foundation of that credibility, Rückert's forecast was innovative because of the introduction of the element of population composition as an exogenous factor in his calculation schemes. Rückert was not interested in the age-sex structure of the population, but in population composition by income groups. Like his mentor Valckenier Kips, he was of the opinion that information on the future composition of the population was necessary before a general extension plan could be made. Whether a town extension had to accommodate a population consisting mainly of the well to do or the working class made a difference.

Rückert thought it unfeasible to distinguish social class, but he saw ways of classifying the population according to income. Once such a distribution of the population was available and the housing demands of each income group were known, it was relatively easy to establish the kinds of dwellings and in what quantities the extension plan had to provide. He used taxation data and poll-tax figures for 1889, 1899, 1907 and 1912. However, analysis of the data of six income groups did not allow him to assess a trend of the relative proportions of each income group. More or less arbitrarily, he assessed the income distribution of the population from the data of 1907 and 1912. He was then able to assess a 'normal type dwelling' for each income group and a 'normal type dwelling lot'.

On arriving at this stage of the socio-economic foundation of his plan, he aborted the attempt to indicate housing districts separately for each income

94 A similar reserve was demonstrated by Angenot (1934) but because of different arguments. He abandoned the idea of finding an explanation of the observed tendency of natality, because it would mean that he had to enter the field of sociology, where sociologists were still debating the issue of natality decline.

group. His argument was that setting up a plan as one wished was easy; anything may be put on paper. Practice, however, ridicules all possible paper divisions. One can project a working class district, but if the flow of workers does not direct itself there, it may become a luxury district instead. The reverse may also occur (Rückert, 1917, p. 61).

It was 15 years before there was another general extension plan with socio-economic and demographic foundations of comparable quality. Rückert himself was convinced that he had turned out good work: *'As far as I know .. our plan has been the first one in the country that was based totally on data, derived from the development of the town; for that reason it is said that it is a stage in the history of Dutch town planning'.*[95] His plan was well received by his colleagues and he was requested to show it repeatedly. Roegholt adopted Rückert's outlines for the social scientific foundation of a general extension plan for the foundation of the plans of metropolitan regions (Roegholt, 1925).

De Casseres' standard work on town planning theory (1926) advocated Rückert's method of working. His report was praised as late as 1932 because of its *'partially excellent explications of the significance of economic research for town planning questions'* (Grondslagen, 1932, p. 91, n. 2).

Towards a standardisation of municipal forecasting

After Rückert's report of 1917, little took place that could stand comparison with his plan. The next decade saw considerable extension plan activity, to such an extent that the Standing Committee for Extension Plans of the Province of North Holland decided that directives for the evaluation of extension plans by the provincial authorities were needed. Once the directives were available, it became clear that they could be used for the instruction of municipalities. They were published in the form of a Directive for the composition of extension plans (Leidraad, 1928). It can be seen as the first comprehensive town planning manual in Dutch, although not intended as such (Van Kesteren, 1984; Van der Heiden, 1990). The Directives formulated the principles of town planning on which Dutch town planners were agreed. These principles, although largely based on international town planning literature, were rapidly becoming familiar through such activities as Valckenier Kips' town planning courses, Rückert's report, the International Town Planning Conference, the activities of NIVS, and Roegholt's thesis.

---

95 Municipal Archives of Tilburg, P.W. 1916 nr. 16, in: Van den Brandhof, p. 52.

The Directive advised that, before the actual plan was made, a comprehensive survey of the present situation of the municipality should be made, followed by a picture of the presumed future development of the municipality (Leidraad, 1928). Socio-economic research would provide the data for the forecast which would result in the list of requirements for the plan (Van der Heiden, 1990).

As to the specifics of population forecasting, the Directive is clear. First, the increase of the population in recent years must be examined, differentiated by natural growth (the surplus of births over deaths) and migration surplus. The size of the migration surplus is significant for the economic condition of the municipality, because as a rule immigration arises when firms or enterprises flourish and the municipality is attractive as a place of residence and location. As many data as possible should be collected with respect to the social class composition of the population. A clear picture of the economic structure is of paramount importance, as economic situation clearly has a great influence on future needs and demands. The mobility of firms should be attended tc. Predictions of the future can never be made with absolute certainty, but information can be acquired which can form the basis of a justified notion of the presumed future development (Leidraad, 1928).

At the end of the 1920s, demographic forecasting came to be applied at the municipal level. It happened in an inconspicuous way, albeit in a rudimentary form. It had the composite components of growth, natural growth and migration surplus, not the true components of demographic growth. The focus on migration is logical from the point of view of Geddes' *'survey before plan'* approach and the focus on the future economic development of a municipality. Apparently without discussion or opposition, the new approach was presented as the standard for local population forecasting. The notion of working with total population growth, geometrical or logistic growth methodology had completely disappeared. Neither in contemporary literature, nor in the archives of the Standing Committee, are traces of opposition to or even debate on the introduction of the proposed new approach to municipal forecasting to be detected. If there were ever signs of a paradigm shift in forecasting, then they were definitely not to be found among Dutch town planners involved in preliminary town planning research of the inter-war period.

Vignette 13: Rückert's forecast of Tilburg (1917)

According to contemporary town plannng theory, a general extension plan had to provide for a future population twice the size of the present population. The most obvious procedure in Tilburg was to start from a future population of 112,000 inhabitants. By dividing this number by the average population density per hectare, the required plan area could be calculated. Rückert, however, preferred another approach; starting from a plan area of the size that in all likelihood would be built on, and comparing the resulting population density with the figure considered desirable. The resulting plan was 337,53 hectares for a population of 155,878 with a population density of 130 inhabitants per hectare, a figure conforming with the current norm.

The plan was checked by calculating the population size that would result from dividing the average frontage (in meters) per inhabitant in the existing built-up area into the total number of running meters of frontage in the extension plan. Rückert arrived at a total of 157,774 inhabitants, close to the outcome of the first method. Rückert then estimated the average growth rate per year of the future population from the observed population growth rates of the periods 1879-1909, 1899-1909 and 1909-1914. He expected an assumption for the future growth rate equal to that of the last 'normal' observation period of 1909-1914 would err on the side of caution. With the geometric growth approach, he calculated when the size of Tilburg's population would be doubled (in 1951) and when the planned capacity would be attained (in 1969).

Finally, Rückert checked the plausibility of the assumptions made with an analysis of the yearly birth and death rates, the natural growth rates, the crude nuptiality rates and the absolute natural and total population growth of Tilburg in the period 1890-1914. He observed the same decrease in birth and death rates in Tilburg found in the rest of Western Europe and even a tendency for the natural growth rate to decrease. He had no desire to seek any explanation for the decreasing birth rate.

The fall in the death rate was assumed to be caused by the application of various hygienic measures. The tendency for natural growth to decrease did not alarm Rückert, for he was confident it would rise again. The decrease could be explained from fluctuations in the crude nuptiality rate, caused by the military mobilisation in 1914 and the housing shortage then characteristic of Tilburg. Since local authorities were seeking to reduce the housing shortage, Rückert assumed that the nuptiality rate would rise again, leading thereby in due course to an increase in the birth rate.

Rückert concluded from an analysis of the observed time series that natural growth rather than an immigration surplus had brought about the population increase in Tilburg. The construction of the Wilhelmina Canal and alleviation of the housing shortage could have resulted in improvements in the economic conditions and attraction of Tilburg and hence to an increase in immigration. Through their young average age, migrants could also increase the birth rate. His forecast was therefore to be taken as a minimum forecast.

7.7 | A Theoretical Framework

The underlying principle of the Directives was that a careful analysis of the socio-economic and demographic situation (the survey) would provide the data for a forecast as the basis for the assessment of the list of requirements for the plan. Once the requirements were known the actual plan could be designed. It was a matter of discussion whether the plan (the design) was the logical outcome of the list of requirements resulting from the survey, or whether there was something in between, a gap that could only be bridged with the intuitive knowledge of the planner/designer (formerly the architect). Quite a few of the advocates of the *survey before plan* or preliminary town planning research approach, whether of town planning, geographic or sociographic origin, had difficulty in accepting such a gap. They were under the spell of 19th century positivism, putting too much faith in knowledge (Faludi and Van der Valk, 1994, p. xiii). The idea of urban and regional planning as an element in a decision making process was still far beyond them.

The uneasy relationship between the domains of knowledge of the future and its construction was analysed by one of the most creative theoreticians of town planning of the Netherlands in the inter-war period, the engineer J.M. de Casseres. De Casseres was an outsider. He was one of the few Dutch town planners of that period who did not graduate from Delft Technical University, receiving his education and experience in Germany, France and England.[96] In 1924, aged 22 years, he became an associate member of the British Town Planning Institute. Two years later he published a book on town planning (De Casseres, 1926). It was an attempt to integrate the sciences that touch upon the spatial designing of town and country –sociology, geography, psychology, statistics and law– into the new science of 'planology' (Bosma, 1993).

---

[96] Joël Meijer de Casseres (1902-1990) has only recently been recognised as one of the most original Dutch theoreticians of town planning of the inter-war period and an advocate of scientific town planning. He got his training in Dresden, Paris and London. In London De Casseres was acquainted with Geddes' surveying work, among others through the town planning lectures of Patrick Abercrombie, who signed for the foreword of his book on town planning (1925). In Dresden he worked as an architect at the Department of Town Extension in 1923. His work in Dresden brought him his first experiences with the design of extension plans from a regional perspective. In the world of contemporary Dutch town planners he held an isolated position (De Ruijter, 1979; Stolzenburg, 1984; Bosma, 1990, 1993).

In De Casseres view, town planning is a social activity, rooted in society. He opposed the view of mainstream town planning, the view of the 'designers', that town planning was essentially founded on artistic, not sociological considerations. Town planning rests on three important social phenomena: Population movement, social relationships and economic conditions. Population movement exerts the biggest determining influence on town planning; it is the most natural and at the same time the least complicated factor in town planning activity. Future demands and provisions like new housing are closely related to future population. Town planning is the instrument for constructing the material structure that forms the basis of social life (De Casseres, 1926).

Population forecasts may determine how many people there will be, but give no answer to the questions how they will work or live. Past population increase can be expressed in quantitative terms, but, for the future, things are different. It is the difficult and delicate task of town planners to design a spatial structure in which future generations are offered the opportunity of a full flowering of work and life. Knowledge of the dynamics of society are therefore more important than its current state. The town planner has to survey the future, but is no better equipped than the sociologist when seeking assurance that the prophecy will come true. The sole 'delicate' instrument available is the careful investigation of the physical and psychic characteristics of the municipality and of traffic relations, based on and related to a study of present living and working conditions.

De Casseres was outspoken about the extrapolations made by mathematicians. In his view such extrapolations had little to do with the future reality the town planner has to deal with. Although it is not clear what kind of extrapolations De Casseres was referring to, it is likely that he meant logistic growth extrapolations. His book was published in 1926 but the draft was finished in the summer of 1925. It is possible but not probable that he was referring to the extrapolations of Rückert (1917) or Methorst (1922), for neither of them was a mathematician and the extrapolations they made can hardly be called mathematical. It is unlikely that he had knowledge of Wiebols' book (1925). Given the fact that his frame of reference was town planning in England, based on his experiences in England and on mainly English literature it is highly probable that he referred to extrapolations of English mathematicians. In that case the forecast of Bowley or Yule's article of 1925 and the debate on the merits of the logistic growth methodology at the 1925 meeting of the Royal

Statistical Society where Bowley and Yule crossed swords may be close to the mark (see Chapter 4).

From references in his book it is clear that De Casseres was familiar with the work of Yule and Bowley and the School of Mathematical Statistics. He mentioned for instance, that in England mathematics were readily used and taken as an instrument for the determination of social and spiritual moments that are most difficult to tackle otherwise; there is a search for objective results by the mathematical operation of statistical data. Mathematical statistics, however, only attain a mathematical truth, which rarely coincides with real likelihood. Statistics no more than sociology and town planning has a neutral, in the sense of an indisputable, validity. With satisfaction De Casseres referred to the characterisation of English mathematical statistics by the German statistician Franz Zizek as "*a specific school of the doctrine of the mathematical method, that rests above all on the application of the calculus of probabilities and error theory. Its significance is theoretic rather than practical*" (De Casseres, 1926, p. 258).

This is not to say that De Casseres wanted to diminish the importance of mathematical extrapolations. He thought highly of (mathematical) statistics, for there the town planner finds the most important instrument for surveying the future. On the other hand he refused to overestimate statistics' significance and he thought it regrettable that most town planners, as town planning literature seemed to show, lacked insight in the very substance and significance of statistics. Too often, it is forgotten that statistical data has to be fertilised by human intellect in order to obtain a logical meaning. Statistics, no less than sociology, is derived from '*Geisteswissenschaft*' (humanities), which spreads a conceptual web between the observer and reality. The perception of reality depends on the nature of the web. It is important, therefore, to know how statistical data have been collected and time series established.

Particularly noteworthy is De Casseres' suggestion to enter an unknown quantity in (statistical mathematical) calculations of the future; a quantity that can influence the results arbitrarily, because it can take on any value (De Casseres, 1926, p. 12-13). The idea was promising, but unfortunately De Casseres did not elaborate on it. The notion of such a parameter could mean that future development was not a mere extrapolation of past trends; it could take different directions, for instance because mankind has a vision of the future hoped for or believed in; also because mankind is to a certain degree able to design (construct) a future that is wished for, or because of the influence of accidents on future developments. In that case, it is the insertion of the dilemma of positivistic (social) engineers of the inter-war period, namely that the design of the future does not follow directly from a socio-economic and demographic survey and forecast, in mathematical extrapolations and quantitative forecasts. It would also mean the translation into mathematics of the gap by which social survey and social forecast are separated from a plan's actual design. The parameter would represent the creative intuition of the town planner-designer and it would be De Casseres' way of preserving the unity of the town planning activity, consisting of positivist research and creative intuitive design.

Whatever the truth of the interpretation, it is clear that De Casseres and Thomas Adams, originator of the regional survey of New York and environs in 1929 (see Chapter 4) held opposite opinions. While Adams was seeking to eliminate the forecaster from the system to be forecast, De Casseres put him in the very centre. At the end of the 1920s, the belief in a law of population growth and the acceptance of demographic forecasting as a speculative approach to the future of population coexisted. The antithetical positions of Adams and De Casseres illuminate the drastic changes that foreshadowed the arrival of demographic forecasting in urban and regional planning. The shift of paradigm in population forecasting had far reaching consequences for the position of the forecaster, both in the forecasting process and in society.

The consequences are discussed more elaborately in the next chapter. Society, whether in the disguise of a town planner, sociologist, or mathematical statistician, is the agent constructing a conceptual network between the observer and (future) reality which blurs the view of (future) reality. Future reality cannot be caught in a law in the form of a mathematical equation, unless a parameter of an unknown quantity is inserted. By stressing the necessity of surveying the dynamics instead of the states, De Casseres underlined the significance of forecasts without overvaluing their relevance.

In the mid 1920s, De Casseres provided a theoretical philosophical framework for contemporary presumptions, expounded earlier in Rückert's report and the Directive of the Standing Committee, that in all likelihood there would be a difference between the future that was going to be and the calculated future; a difference that not even the application of mathematics could avert.

7.8 | The Propagation of Demographic Forecasting in Planning

By the end of the 1920s, the traditional forecasting of total population was being almost silently substituted by the demographic component approach. The 1930s was the decade of the true breakthrough of the demographic forecasting methodology to urban and regional levels. Incidentally, and in an indirect way, modern demographic forecasting had already made its appearance on the stage of local forecasting when Hudig, as a member of the Standing Committee for the Extension Plans in North Holland, made a forecast of the future population of Amsterdam based on the assumption of a constant proportion of Amsterdam in the future population growth of the Netherlands as calculated by Wiebols (Grondslagen, 1932). In doing this, Hudig had circumvented the problem of a separate estimation of future migration. For a thorough forecast of the population of a big town Hudig's solution was not sophisticated enough, partly because the influence of the migration factor remained unknown.

The first, and by far the most influential, modern forecast was that of Amsterdam in 1932, and was as innovative as that of Rotterdam (1934). Part of the innovative contribution is discussed in Chapter 5. The almost simultaneous appearance of the forecasting studies of Amsterdam, The Hague and Rotterdam, each in their own way impressive pieces of work, raises the suspicion that their publication was based on mutual agreement of the municipalities, or the municipal forecasters, involved[97]. One might have expected a substantial exchange of ideas, solutions to methodological problems or other information among the forecasters involved. Surprisingly, the opposite appears to have been the case. Bakker Schut was surprised that the Amsterdam report was published just a few months before he had finished his own study of The Hague, as if he were ignorant of the endeavours of his Amsterdam counterparts (Bakker Schut, 1933, Foreword).[98] Apart from a few minor remarks of a non

---

97 The publication of these studies was not a self-evident matter. The Amsterdam study was financed by the municipality; the expenses of Bakker Schut's book were paid by himself; the mayor and alderman of The Hague only gave consent to do so (Bakker Schut, 1933, foreword). The study by Angenot was published by a private organisation of interested citizens of Rotterdam, the society for Town Improvement *Nieuw Rotterdam* (Angenot, 1973).

98 The Director was Engineer Piet Bakker Schut, not to be mistaken for his son, Frits Bakker Schut who was to become the first director of the National Physical Planning Agency.

methodological character, Angenot refrained from referring to the studies of his colleagues in Amsterdam and The Hague.

The studies of Bakker Schut and Angenot presumably had a common origin, a discussion at a meeting of the Union of Directors of Public Works in April, 1932. Bakker Schut had focused attention on the change of direction of national population growth and its consequences for the development of Dutch towns. The list of references in his book makes it clear that he was well informed about contemporary national and international demographic literature on the population issue and the declining birth rate, among others about Kuczynski's work on the balance of births and deaths. Bakker Schut's statement at the meeting was a reflection of the (inter)national debate: Population development had arrived at a turning point, an end to further growth was impending, the national population would have reached its maximum before the end of the century and the possibility of a future population decline was not to be excluded. For that reason, the time had come for municipal planners to investigate the consequences for the big cities (Bakker Schut, 1933). Urban and regional development was seen as the derivative of the national population issue, which explains why both Bakker Schut and Angenot started from a national forecast before they proceeded to forecast the population of their respective municipalities. At the time of the meeting where Bakker Schut made his statement, the Amsterdam forecast must have been at an advanced stage. In contrast with the studies of Bakker Schut and Angenot, it was part of a true preliminary town planning investigation that had to serve the purposes of a general extension plan.

7.9 | The Population Forecast of Amsterdam, 1930-2000 (1932)

The forecast of population and housing need in the report of the Garden City Commission, discussed in Chapter 6, was the outcome of action taken by the municipal Department of Housing. It was already out of date at the moment of its publication. In the meantime, the town council decided that the future growth of the population of Amsterdam had to be housed in extensions of the town and not in garden cities at a distance. For that reason, it was decided that a General Extension Plan should be made under the auspices of the Department of Public Works, which had created a special division of Town Development for that purpose. In line with the principles of modern town planning theory and the Directive of the Standing Committee for Extension Plans, it was decided that the plan had to be based on an extensive survey and that a population forecast should be undertaken first.

The General Extension Plan of Amsterdam of 1934 was the first comprehensive pre-war Dutch example of modern town planning founded on scientific principles. Van Lohuizen's name is closely associated with the preliminary town planning research part of the plan. The feasibility of the ultimate design was to depend on scientific arguments derived from a comprehensive survey. A central part of the scientific foundation was the population forecast of 1932. It was made by Delfgaauw, but under Van Lohuizen's supervision.

In justifying their forecasting method, Van Lohuizen and Delfgaauw felt obliged to explain that it was no longer sufficient to be satisfied with the application of Valckenier Kip's simple rule of almost a quarter of a century before that an extension plan should provide for double the number of inhabitants present at the moment of its adoption (Grondslagen, 1932). The influence of Valckenier Kips was far reaching indeed. Neither did they consider it justifiable to be satisfied with calculations like those of the Garden City Commission, which all had in common the simple extrapolation into the future of the observed past growth rate of the population.

The arguments of the Garden City Report for not applying Wiebols' demographic method had been countered in the meantime: Wiebols' method was recognised as the best by a majority of experts; the results of methods based on more sophisticated demographic analysis were considerable,compared with mere extrapolation of growth rates; the magnitude of the task might not be an obstacle to the aim of the forecast (Grondslagen, 1932).

The forecasters were faced with the dilemma of whether to opt for an economic or a demographic approach. The economic approach, which was preferred as is discussed in Chapter 2, had to be dismissed as a serious option because it was complex and would have been too time consuming; according to the forecasters, it would have implied the assessment of the population capacity of the country, a factor largely determined by mondial economy (Grondslagen, 1932). In theory, the procedure to be followed would have been the determination of Amsterdam's future economic position within the national economy and its attractiveness as a residential town. Starting from the expected total of the national population in the year 2000, already calculated by other forecasters, one could try to determine what proportion of national population

Vignette 14: Theodoor Karel van Lohuizen (1890-1956)

Van Lohuizen started his studies at both Leiden University and Delft University of Technology (1909). In 1910 he decided to concentrate on civil engineering in Delft, specialising in road building and hydraulic engineering. While a student at Delft and until 1935 he participated in the activities of the Social Technical Society of Democratic Engineers and Architects, a society aimed to arouse the interest of future engineers in the social aspects of engineering. Through his participation in the society, he became acquainted with engineers who, like him, would become pioneers in housing and town planning (e.g. P. Bakker Schut; M.J. Granpré Molière; A. Keppler) and he became involved in the network of the Netherlands Institute of Housing and Town Planning.

At Delft University, Van Lohuizen became acquainted with town planning through the courses of professor J.H. Valckenier Kips. Here he learned the importance of insight into the future development of population and its composition as a foundation for a municipal extension plan. Van Lohuizen's main interests were in preliminary town planning research, the social-economic and infrastructural foundations of a plan. For much of his professional career he held the rather positivist view that the actual plan would follow the findings of scientific research in a logical, rational way.

Van Lohuizen began his career as an engineer at the Department of Housing of Rotterdam, where he stayed until 1928. His work at the Housing Department involved him closely in the estimation of changes in the number of households and in the calculation of (future) housing need. In his Rotterdam period he developed an innovative, but complex and rather time-consuming method for estimating housing need. He was involved in the extension of the town, the improvement of the inner city and the survey for a regional plan of the Rotterdam and the Hague area. His work on this survey resulted in an exhibition of maps, graphs and diagrams at the *International Town Planning Conference* of 1924 at Amsterdam, where the central theme was the necessity of regional planning. Van Lohuizen's graphics were designed to convince politicians and decision makers of the urgency of planning the western part of the province of South Holland.

In 1928 he was appointed as a researcher in the new Division of Town Development of the Department of Public Works in Amsterdam, where he was responsible for the demographic and socio-economic foundations of the general extension plan. His work on the survey of the Rotterdam-

The Hague area and at the Division of Town Development of Amsterdam made him into a well respected expert, both in town planning research and population forecasting. When the Netherlands Central Bureau of Statistics embarked on national population forecasting in the 1940s he became an independent advisor (Source: De Gans, 1997).

increase would find employment in Amsterdam, given the development of its economic potentialities, and what proportions of those employed would actually live within the municipal boundaries.

As is demonstrated below, the idea of an economic approach was never completely abandoned during the 1930s. Although from a theoretical point of view the demographic method was considered to be second best, it was decided to use it nevertheless for reasons of practicality. The method was based on the assumption of an absence of unexpected changes in the factors of population increase in the future. Furthermore, it was assumed that the proposed demographic forecast was a minimum forecast. The point of departure was to make a forecast of a future population that would be realised, even in the case of a less favourable economic development of Amsterdam (Grondslagen, 1932).

The forecasters endorsed Wiebols' opinion that a forecast can only rest on the assumption of a future continuation of a development that had already been persistent for some time. They shared Wiebols' opinion that this did not necessarily mean that observed past developments would continue into the future with certainty: The future lay in the dark; all pronouncements about the future were speculative. In the view of the forecasters, the main purpose of such a calculation was to measure the impact of forces, which were operating already in the present. If other forces arose, these would have to be accounted for later (Grondslagen, 1932, p. 99, n. 13).

Why Wiebols' method was preferred to 't Hooft's cohort approach is extensively argued. Why 't Hooft's standardised mortality approach was repudiated, is explained on the basis of a thorough analysis and an account of the arguments used in the debate on the merits of that method (Grondslagen, 1932; also Chapter 5). In Chapter 2 it is related that Delfgaauw, while he was working at the National Institute of Housing and Town Planning, became acquainted with the recent performances of the forecasters of the population of the Netherlands. It may have helped him form his view about the best way to make a municipal population forecast. It is not unlikely that the preference for the Wiebols method was also influenced by Van Zanten's reviews of 't Hooft's book of 1929 (Grondslagen, 1932, p. 96 n.7). Van Zanten (1931) appreciated that 't Hooft had brought to an end the idea of geometric population growth. He appreciated also his cohort survival-in-the-Netherlands schemes. Van Zanten preferred Wiebols' approach of the calculation of the future number of births with fertility rates, which he considered much more refined than that of 't Hooft (De Gans, 1994a).

Following Wiebols (1925), only the female population was forecast while similar assumptions of future age specific probabilities of death were used. Like Wiebols, the future number of (female) births per year in a ten year forecasting interval were calculated by relating average general fertility rates (age group 15.5-49.5) to the number of women in that fertile age group at the beginning of the interval. In contrast with Wiebols, it was decided to take the general fertility rate of London (in 1925) instead of France (1910/1911) as the limit of future general fertility of Amsterdam, because general fertility of Amsterdam was already below the level of France in 1910/1911.

The elaborate calculation procedure of the forecast was conveniently organised and resulted in a population of about 450,000 women in 1990 and a total of less than 900,000. Seven other forecasts were made based on alternative assumptions with respect to future migration.

Generally speaking, Van Lohuizen and Delfgaauw's forecast of Amsterdam was exemplary in its thorough account of each individual decision and careful critical overview of contemporary population forecasting. As the first to apply Wiebols' modern demographic method to urban planning and forecasting, it was an innovative and pioneering piece of work. This does not mean, however, that the underlying assumptions were trailblazing. From a modern point of view some of them seem surprisingly naive, particularly those with respect to migration: The forecasters assumed that the age-specific emigration and immigration proportions per ten year interval of the forecasting period would be equal to those of the period 1900-1920.[99]

---

99 A problem the forecasters had to face was that on January 1, 1921 the neighbouring municipalities of Buiksloot, Nieuwendam, Ransdorp, Sloten and Watergraafsmeer had been added to the municipality of Amsterdam. In contrast with Amsterdam only total immigration and emigration figures, not specified by age, were available (Grondslagen,

On the other hand, in terms of modelling the introduction of age specific migration proportions in the main variant of the calculation schemes was sophisticated.[100] From an analysis of female migration statistics of the period 1880-1920 (ten year intervals), the forecasters had discovered that age specific migration surpluses could differ considerably, even though the total migration surplus was negligible.

The forecasters were aware that, in case of a balance between emigration and immigration totals, differences in the respective age composition of emigrants and immigrants could have a considerable effect on future population size. For that reason it was thought necessary to extend Wiebols' calculation schemes with age specific migration. The age specific female population at the end of a ten year interval was calculated from the survivors at the end of the interval of the women present in each age group at the beginning of the interval and the age specific migration surplus, with a correction for mortality of the immigrant population.

The calculation procedure is reconstructed in the form of the following demographic equation:

$$P_{x+1;t+10} = P_{x;t} * (1-{}_{10}q_x) + M_{x;t-t+10} * (1-{}_{10}q_x)$$

wherein

$$M_{x,t-t+10} = P_{x;t} * m_x : 1000$$

and

$P_{x,t}$ The female population in a five year age group (x) at the beginning of the ten year interval;

$P_{x+2;t+10}$ idem for the next older five year age group ($x+1$) at the end of the interval;[101]

$_{10}q_x$ the ten year probability of death of women of age group x at the beginning of the interval and of the immigrants of that age group;

$M_{x;t-t+10}$ the migration surplus of women of age group x during the ten year interval;

1932).

100 Age specific female migration proportions were calculated by relating the female migration surplus of an age group per ten year interval to the number of females of that age group at the beginning of the interval. In other variants only total future migration was taken into account.

101 The calculation procedure is explicated in Grondslagen (1932, p. 64-79, schemes Xa-Xh).

m_x the migration surplus proportion per 1000 of age group x.

The youngest age group at the end of an interval was forecast with a similar calculation procedure, but this time specified by birth cohort of one year and five year intervals, to the end of the five year interval (age group 0-4).

Reception
The assumptions of the forecast were not readily accepted by all reviewers. It is interesting to note the differences in the remarks made by politicians and by others. Members of the municipal council appear to be alone in arguing that a future emigration surplus might occur as well (Bruyn, 1976). Most critics, however, referred not to migration, but to other aspects of the forecast. Willekes Macdonald, an engineer, wrote a short but very critical review in *De Ingenieur*, arguing that the department of Public Works was arbitrary in its assumption making. For instance, why had the forecasters not opted for the figure of Paris as the limit level of the general fertility rate, which was much lower than that of London. Willekes Macdonald spoke of *'utopian asymptotes to which the curves more or less connected to'* and he criticized the Department for not having made economic forecasts (Willekes Macdonald, 1933; in Hellinga and De Ruijter, 1985). Willekes Macdonald assumed that a population total of 600,000 in 1990 would be much more likely than the calculated total of 900,000. The Director of Public Works riposted that the character and structure of the Amsterdam population was more similar to that of London than that of Paris and that the range of demographic development of a population lay between rather narrow boundaries. The prediction of the future of industrial enterprise is less solid (De Graaf, in Hellinga and De Ruijter, 1985).

Van Zanten commented in a letter to the alderman responsible for statistics that a prediction that the population of Amsterdam would no longer increase from the year 2000 on would be received with some laughter in the next century (Bruyn, 1976: p. 45). That he was ambivalent in his judgement of forecasts like that of Amsterdam has been discussed before: In terms of a scientific piece of work he thought that Amsterdam could consider itself lucky that this report had been produced. As to the practical use of this time consuming calculation project for planning purposes, he was less satisfied. Authorities could easily be misled by them, because of their speculative nature. He would not dare to rank it higher in his appreciation than a simple qualitative assumption (De Gans, 1994b).

In an internal report of the Department of Public Works of Haarlem of 1936, the engineer F. Ottevangers made a noteworthy comment on a technical aspect of the forecast. Both Wiebols and the Amsterdam forecasters had related the general fertility rate of women of age group 15-49 to women of that age group present at the beginning of a forecasting interval. The calculation model, therefore, omitted to account for the fertility of women of age 10-14 at the beginning of the interval, who would be of age 20-25 at the end of the interval. In an implicit way, he advocated a cohort based occurrence-exposure rate approach to the calculation of the future number of births.

7.10 | The Forecasts of The Hague and Rotterdam

In Bakker Schut's forecast for The Hague (1933), migration was treated much more simply. His forecast resembles that of Hudig. The future population size was calculated from two separate forecasts: Future population size resulting from natural growth, and a forecast of future immigration surplus. Future population totals for The Hague were taken as a proportion of the national population as calculated by Wiebols, 't Hooft and Bakker Schut himself (see Chapter 5), starting from three different periods of observation: 1879-1930; 1899-1930 and 1920-1930 (nine variants in total). The future immigration surplus was estimated from two different assumptions: A gradual decrease of the observed immigration surplus to a level of zero in the year 2000; a continuation of the immigration surplus at the level observed in the period 1920-1930 (the last mentioned assumption was thought to be purely speculative). By adding the outcomes of the 'natural growth' forecasts and the immigration surplus forecasts, he obtained eighteen forecasts from which he selected a maximum, a minimum and a medium forecast.

Angenot was convinced of the innovative character of his 1934 forecast of the Netherlands and the Rotterdam Harbour Area. He attributed it above all to the use of age/cohort specific fertility rates for the calculation of the number of births per interval instead of using general fertility, as Wiebols and Van Lohuizen and Delfgaauw had done. Angenot was over modest; he greatly underestimated his innovative modelling power. In terms of formal modelling, he was the most prolific and innovative forecaster in the Netherlands and perhaps even in the international forecasting field. His 'synthetic cohort survival scheme' approach, is discussed in Chapter 5. His formal model of the estimation of future cohort specific fertility rates is referred to in Chapter 4.

Migration has a two-way impact on population development: First, through mere numbers; second, because its age-specific character affects the fertile age categories and thereby the number of births. For that reason, Angenot had to face the problem not only of the estimation of future migration totals as such, but also that of the future age structure of the migrant population and its influence on the future size and structure of the municipal population. He opted for a formal modelling approach, searching for arguments allowing him to simplify his model. The arguments for the simplification came from an analysis of recent trends of migration. For the distribution of estimates of future migration surplus totals over the resulting age groups (the groups to be reckoned with), he worked with (constant) yearly age-specific immigration surplus proportions.

As far as is known, Angenot was the first Dutch forecaster to introduce a matrix notation and matrix mathematics into Dutch population forecasting; the results were presented in the form of matrices for the male and female immigration surplus population respectively (Angenot, 1934). In principle, his simplified matrix approach was apt to further specification for both national and local and regional forecasting purposes; it comes close to Leslie's matrix approach, although in the presentation it lacks the clarity and transparency of Leslie's model.[102]

Angenot's formal forecasting model was presumably too time consuming to find much following and too mathematical to be readily applied by the new generation of preliminary town planning research professionals, the social geographers and sociographers. Angenot's friend Bijhouwer, however, immediately understood its wider implication. He wrote him: *'Dear Lo, I have ploughed through your publication with an awful lot of pleasure; one of the nice things in it seems to be your way of generalising (....) which allows for substitution to one's liking. Here and there it becomes devilish complicated; for that reason I tend to see demography as demonography'*.[103]

102 Excerpts of Leslie's 'On the use of matrices in certain population mathematics' in *Biometrika* 33, 1945 are presented in: Smith and Keyfitz, 1977, pp. 227-254).

103 Letter engineer J.T. P. Bijhouwer to Angenot d.d. 29-1-1935 (NAi, Arch. Angenot, inv. nr. 5).

7.11 | International Reception

Forecasting methodology developed in the Netherlands independently of foreign influences to a high degree. This holds for national population forecasting, with the possible exception of Methorst's forecast of 1922, which may have been influenced by Fahlbeck, and his forecast of 1937, in which Bürgdörfer's *'Bereinigten Lebensbilanz'* approach was applied (see Chapter 4). It holds in particular for urban forecasting. Van Lohuizen, Delfgaauw and Angenot demonstrated that it was possible to apply the Wiebols methodology to the migration component, making it suitable for urban and regional forecasting purposes. The calculation schemes of Van Lohuizen and Delfgaauw are transparent and easy to understand without a mathematical background. It became a model for post-war municipal forecasting.

Angenot's Rotterdam forecast was much more sophisticated in comparison with that of his colleagues in Amsterdam. Not only because he was the first to enter age/cohort specific fertility rates in his calculations, as Wiebols had advocated 1925 but had not yet been able to put into practice, but in particular because of his attempts of formal matrix modelling. Angenot's forecast had the disadvantage that it was less easy to grasp. It lacked the transparency of the forecast of Amsterdam and of the matrix model that was developed in the United States by Leslie in 1945, which became the international standard matrix approach to forecasting after World War II.

Although few direct foreign influences on the development of the new standard forecasting methodology could be discerned, international contacts and exchanges of information on forecasting between town planners have occurred. The municipal forecasters were more interested than Wiebols and 't Hooft in communicating with international colleagues, were better equipped, and had more opportunities to do so. Pioneers in the modernisation of urban forecasting differ from those in national forecasting in that they were part of the international community of planners actively exchanging information about their innovative endeavours.

In the wake of contemporary international interest in the General Extension Plan of Amsterdam, the underlying population forecast also received some attention (e.g. Amsterdam, 1935; Dougill, 1936; Giedion, 1940/1966). The attention paid, however, was mainly descriptive or focused on the calculated future population size of Amsterdam; discussions of the merits of the forecasting method applied were absent. The survey and the plan of Amsterdam, its layout, diagrams, and illustrations were put forward as models to be emulated by other cities. Much of the work was considered to be pioneering in character and its usefulness for planners was correspondingly enhanced (Dougill, 1936, p. 10). There is little doubt that the demographic forecasting work was seen as part of the pioneering character of the plan, but it was not mentioned separately. Giedion (1940/1966, p. 807) merely stated that "*methods of forecasting the future needs and composition of its population were developed very early in Holland.*"[104] There are no methodological references to comparable international municipal or regional forecasts in the Dutch reports of municipal forecasts in the 1930s. In Delfgaauw's contribution to the Journal of the Town Planning Institute on the population forecast of Amsterdam, there was no discussion of the merits of his Amsterdam forecast within the context of contemporary international municipal forecasting. Under the assumption that the innovative Dutch forecasters/town planners of the 1930s were not working in isolation, it can be concluded that urban forecasting in the Netherlands in the 1930s held a pioneer position. There were few forecasts of foreign cities of an equally high methodological quality.

Angenot, like Delfgaauw, took care that his work became known outside the Netherlands by sending copies of his book to colleagues at home and abroad. Rappoport, a town planner of the Ruhrkohlenbezirk of Essen in Germany, commented that it was highly interesting to learn that the foundations and conclusions were the same everywhere: External causes, such as the welfare of the nation, or even relationships overseas with colonies determine the eventual growth of the population of Rotterdam, but for the rest he refrained from discussing the merits of the forecasting method.[105]

---

104 Siegfried Giedion, a German of Swiss abstraction, was very enthusiastic of town planning in the Netherlands. Delfgaauw recognised the importance of contributing to international journals as well as Van Lohuizen. Van Lohuizen may even have stimulated him to do so. Only one year after the publication of the forecast of Amsterdam, Delfgaauw contributed a paper in the English *Journal of the Town Planning Institute*, in which he presented the outlines of the discussion on the forecasting method best to be used, the outlines of the forecasting method opted for and the main results (Delfgaauw, 1933C).

105 In a letter to Angenot d.d. 11 August 1935 (NAi, Arch. Angenot, inv. nr. 6). Rappoport criticised Angenot for not having been clear enough as to the relativity of his forecast of one million inhabitants of Rotterdam to be reached in 1970. He did not think it sensible to reckon with such a figure; he thought it too specific to serve as a base for measures to be taken. It would do to make reservations of the area needed if the forecast eventually might come true.

Insofar as an international discourse on urban and regional population forecasting methodology took place, it was a discourse among planners. An exchange of information on advancements and innovations in the field, as in the Netherlands, remained restricted to the planning profession. The international communities of planners and of statisticians/students of population were separate worlds. Dissemination of information on the state of the art outside the planning profession was not sought. And the international (town)planning community was interested in the design of (extension) plans in the first place, in the demographic and socio-economic foundations of the plans in the second place, and in the sophistication and details of the founding demographic forecasts only in the third and last place. This affords a further explanation of why the exploits of the Dutch innovators of municipal forecasting methodology of the 1930 failed to figure in the international intellectual histories of modern population forecasting.

7.12 | A New Standard But a Challenged One

The context of municipal and regional population forecasting in the inter-war period was totally different from national demographic forecasting. National population forecasts of the period served purely scientific purposes closely related to interest in the population issue: The danger of overpopulation in the 1920s and the fear of population decline in the 1930s. There was no direct connection with government policies. Municipal forecasting, on the other hand, as the core of preliminary town planning research, was directly related to municipal housing, planning and policy purposes. In this connection, Angenot's introduction to his forecast of the Netherlands and of Rotterdam is revealing.

The reader of his book is warned that the result of his calculations, 11 million inhabitants in the Netherlands and one million in Rotterdam, was not a prediction, but *'a rational expectation, that has to serve as a guideline for action. Each future oriented action should be judged by the circumstances, that are thought to be the most probable, given the current state of our knowledge'* (Angenot, 1934, p. 8) The device of his book is a statement from John Meynard Keynes' *A treatise on Probability* (1921): "*The probable is the*

Vignette 15: Professor G.Th.J. Delfgaauw (1905-1984)

At the age of 23, Gerardus Theodorus Jozef Delfgaauw wrote: *'This part of the investigation is of course open to further elaboration and we are fully convinced of the limitations of our contribution to the question of the future development of our big towns; to some extent it may dispose our minds to joy that we do not arrive at a well-formulated prediction, for the fate of most societal prophecies is far from enviable. On the other hand, however, it is in many cases impossible to take measures today when one has not the slightest idea of what tomorrow will bring, so that even the smallest ray of light in the dark of the future is accepted in gratitude. We hope that our contribution will be considered as such'* (Delfgaauw, 1932, pp. 9-10). Delfgaauw, the man who was to apply the Wiebols method to an urban population, was already capable of seeing the relativity of all acts of societal forecasting.

In the story of inter-war innovation in urban and regional population forecasting, the names of Van Lohuizen and Delfgaauw are closely linked. They are irrevocably connected to the study of the demographic foundations of the General Extension Plan of Amsterdam of 1935 and to new ways of regional population forecasting.

Delfgaauw had become familiar with the literature on population while working at the Netherlands Institute of Housing and Town Planning (NIVS) in 1928, and with the practice of population forecasting at the Division of Town Planning of Amster dam. He soon became a forecasting expert; the population forecast of 1932 was mainly his achievement.

In 1934, Delfgaauw obtained his PhD at the Municipal University of Amsterdam on municipal land politics, a subject closely connected with one of the other foundation studies of the General Extension Plan. Born in Amsterdam in 1905, Delfgaauw took his school leaving examinations in 1922 and, in the very year was established, entered the Faculty of Commercial Sciences (later the Faculty of Economic Sciences) of the Municipal University of Amsterdam to study Economics. He belonged to the first generation of economists to graduate from the faculty in 1927. Professor Frijda supervised his thesis on housing and rent and introduced Delfgaauw to D. Hudig, Director of NIVS. Delfgaauw's involvement in town planning, demography and population forecasting ended in 1946 when he became professor of Political Economics in the University of Amsterdam (Source: De Gans, 1997).

hypothesis on which it is rational for us to act" and later the device is quoted again but now within a larger context "*Given as our basis what knowledge we actually have, the probable is that which it is rational to believe. This is not a definition. For it is not rational for us to believe that the possible is true; it is only rational to have a probable belief in it or to believe in it in preference to alternative beliefs. To believe one thing* ***in preference*** *to another, as distinct from believing the first true or more probable and the second false or less probable, must have reference to action and must be a loose way of expressing the propriety of* ***acting*** *on one hypothesis rather than on another. We might put it, therefore, that the probable is the hypothesis on which it is rational for us to act.* '(J.M. Keynes, in: Angenot, 1934, p. 8).

Angenot's position with respect to forecasting was totally different from the attitudes of Verrijn Stuart and Van Zanten, who stressed its speculative nature. To Verrijn Stuart and Van Zanten its speculative nature was a reason for eliminating forecasting from the field of statistics. Van Zanten even sought to eliminate all semblance of accuracy in the prediction of the future from policy making by advocating practice-oriented forecasting methodologies that were as simple as possible. To Angenot, forecasts were part of rational decision making (action). To him, at the heart of policy-oriented forecasting, a forecast was a probable demographic future, which meant a probable belief to be put to use in rational action.

In a course of town planning instruction at the University of Technology of Delft, Angenot taught his students that forecasting was not a search after certainty for the future, for that cannot be given. Forecasting was nothing more, and nothing less, than a search for the most likely continuation of observed trends.[106] This search was the core of the forecaster's responsibility. In his view, there was a direct relationship between preliminary town planning research, forecasts and plans, for the more profound our investigations, the more precise our assumptions of the future, the better our plans. There is no talk, as with De Casseres, of a parameter that can take on all values, nor of the dilemma of the artistic intuition that comes in once a plan is being designed. Better plans automatically follow as investigation becomes more profound and precise.[107]

---

[106] Course of town planning instruction 1934 (NAi, arch. Angenot, inv. nr. 314).
[107] This rather positivistic attitude would eventually cost him his leading position in Rotterdam.

Municipal and national demographic forecasters of the inter-war period have in common that they were intent on informing policy makers and society of the most likely course of future population. The belief that the processes of demographic growth could be influenced by policy action appears to be absent. Because urban and regional forecasting is action directed, and because the quality of a plan was thought to depend on forecasts that were as accurate as possible, urban and regional forecasting could not circumvent the problem of dealing with migration. The problem was how to do so.

The dilemma of urban forecasters of the 1930s and later decades was that future population development largely depended on economic development, but it was almost impossible to forecast economic development and its consequences for the labour market. The consequences for the labour market were considered to be the factor determining the number of households that could find a living, and by implication the number of immigrants and commuters. The preferred procedure therefore was a labour market forecast as part of the forecast of the economic development of a municipality. Because this proved both time consuming and uncertain, another line of forecasting had to be followed. This was the line of the mere demographic approach. For a long time to come, the demographic approach was thought to be a second best solution, because it started from the misconception that urban and regional population growth was an autonomous process. It is of course in assumption making with respect to future migration that the socio-economic and demographic approaches overlap, as was clearly stated in a study on the future socio-economic development of Amersfoort, a few years after the Second World War (Klaassen, Van Dongen Torman and Koyck, 1949).

The migration dilemma accounts for the increasing frustration of forecasters of the 1930s and for their attempts to seek support in economic forecasting. In this respect, demographic forecasting in the 1930s does not differ much from the former belief in a law of population growth. The forecaster is still an informer about the course of future population; that he is part of the forecasting process and of the system to be forecast was to a high degree beyond his scope or interest.

Many contributions to the practice of regional forecasting advocated the exploration of indirect approaches that started, as in Hudig's forecast of Amsterdam, from a national population forecast. The most influential contribution came from Van Lohuizen and Delfgaauw (1935). It was merely meant as a first attempt at regional forecasting.[108] The authors were startled when they discovered that what was intended as an experiment had become a standard approach, finding general application to regional forecasting (Angenot, 1973). It originates in Van Lohuizen's ideas regarding the socio-economic foundations of general extension plans and regional plans. In earlier investigations with respect to the development of population distribution in the Western part of the country he had already started from homogeneous socio-economic regions (Van Lohuizen, 1925). Delfgaauw had worked along the same lines at NIVS in 1928 (Delfgaauw, 1932; also Chapter 2).

Like Cannan in 1884 Van Lohuizen and Delfgaauw started from the conviction of the arbitrariness of administrative (municipal) boundaries. They shared Angenot's opinion of the existence of a *communis opinio* that national population would reach a maximum, somewhere between 10 and 13 million inhabitants about the year 2000. They could (migration excepted) build on forecasting methods developed for national forecasting purposes.

Migration, however, depends on economic factors.Urban and regional population growth can only be investigated in relation to that of other towns and regions. The geographic distribution of the national population depends on the relative means of existence of the various regions. A scientifically reliable estimate of future population development based on assumptions of the future economic structure of the country is difficult to make. For the time being indications of the future direction of regional population developments had to suffice.

Van Lohuizen and Delfgaauw divided the Netherlands into 19 separate economic geographical areas, each characterised by the dominant means of existence. The differences in means of existence were the cause of mutual

---

[108] The immediate cause of the regional forecasts of Van Lohuizen and Delfgaauw was a discussion of the question whether the regulation of extension and structure plans of the Housing Act provided for sufficient opportunities to protect natural and scenic beauty against destructive and violating building activities (Bebouwing, 1942: v). Van Lohuizen and Delfgaauw were requested by the NIVS Working Committee for the Day of Monuments (in 1933) to advise on this issue. They started a quest for a simple method of estimating future regional population development. The findings were published in Van Lohuizen and Delfgaauw (1935) and used in the final publication of the working committee (Bebouwing, 1942). Similar but less elaborate approaches are presented in NIVS (1942/1943) and P. and F. Bakker Schut (1944).

differences in population growth. The future population of (combinations of) economic geographical areas was calculated on the basis of the observed differences of population growth between regional populations and the national population and a continuation of the observed proportion of the population of a region in the national population.

The degree to which Van Lohuizen and Delfgaauw were influenced by the 1925 forecast of Wiebols is clear from the fact that it provided the data for the future national population size. It was argued that in this way a maximum population forecast would be obtained, given the decrease in fertility since 1925 (a decrease that was not compensated by a similar decrease of mortality).

As to the future population of the regions distinguished, it was assumed that future population growth would gradually decrease, in line with Wiebols' assumption of the national population. The relative tempo of regional growth as observed in the period 1920-1930 was expected to continue in the forecasting period. Internal migration was expected to decrease linearly with national population growth.

The results with respect to the Amsterdam, Rotterdam and the Hague regions could be compared with the outcomes of the strictly municipal forecasts of Amsterdam and the areas of The Hague and Rotterdam from the first half of the decade. Differences were acceptable and fell within the range of minimax outcomes. Moreover, a retrospective analysis of the proportion of the various regional populations in the total population of the Netherlands had demonstrated that in 14 out of 19 regions the proportions were sufficiently constant. The forecasters concluded that the calculations gave information about the direction of future population development to be expected as long as observed trends of the past ten years continued in the future. For further elaboration, they suggested to take account of specific factors at work in the various regions.

The forecast of Van Lohuizen and Delfgaauw was a first step in the direction of socio-economic forecasting. It did not lessen the general dissatisfaction among forecasters, because of the problem of the accurate prediction of future migration. It was to offer opportunities for the new professionals in regional planning, the economic geographers and sociographers, to contribute to the advancement of forecasting methodology. Van Braam (1948, p. 166) wrote: *'The favoured position of the population forecast in the survey as such does in principle not lead to more reliable, objective and less speculative results, that justify its superiority. The more easy accessibility and treatment of the underlying demographic material at the one hand and at the other hand the lack of sufficient economic data, are the causes of the disproportional increase of the significance of the population forecast in the survey'.*

About the same time, the economists Klaassen *et al.* (1949) published a forecast of the municipality of Amersfoort based on econometric principles, although in a rudimentary form. The sociologist Hofstee earned renown through his forecasts based on a new way of estimating the future number of jobs of a region. It marked the beginning of a new era in which economic geographic forecasting would successfully challenge the dominant position of pure demographic forecasting. The era ended, as far as urban forecasting is concerned, when it became clear that future municipal population development could easier be forecast on the basis of expected extensions of the municipal housing stock (about the 1960s/1970s).

8. The Implications of the New Paradigm

8.1 | The Social Position of the Forecaster

The history of the shift in standard population forecasting methodology has come to an end. By the end of the inter-war period, demographic forecasting had acquired a dominant position. If the history is considered in a strictly Dutch context, there are not many indications of a shift of paradigm in the study of future population. It is true, there was a vehement debate about forecasting method in the Netherlands going on. However, this was a debate on the merits of competing demographic forecasting methodologies, not between demographic forecasting methodology and geometrical or logistic laws of population growth.

The stakes played for in the international debate were different. Here, demographic forecasting had to fight for dominance. In the 1920s, the odds seemed to be against the new methodology and the indications of paradigm shift in the quantitative study of future population are much clearer. Only by taking into account the positions of Dutch participants in international population debates could such indications also be found in the Netherlands.

In this chapter history is left behind. Here, the focus is on a number of closely related implications of the paradigm shift that have generally escaped the attention of the historians of population forecasting. The chapter starts with a juxtaposition of the views of two contemporary town planners, Thomas Adams and Joël M. de Casseres, familiar to the reader from previous chapters. Adams and De Casseres had diametrically opposing views. In his eloquently expressed fascination of logistic growth methodology Adams is the most outspoken representative of the 'old' paradigm. Adams appears to us as a planner who dreamt of finding the law which determines future population growth. The logistic growth law came close to this ideal. Such a law would limit the forecaster's activities to the search for the most adequate mathematical function and the parameters for its most accurate expression. Under such a law, the future would just grow from past and present reality and neither forecaster nor forecasting qualities could act as a watershed between the past and the future.

The view that accompanied the new paradigm was most vividly voiced by Adam's Dutch contemporary De Casseres. Like Adams, De Casseres thought highly of the utility of mathematical growth functions, but his appreciation was of a different nature. In De Casseres' view the image of the future (of future reality) is blurred, because it can only be captured with conceptual networks that are products of the human mind. Mathematical networks do not differ from other conceptual networks in this respect. Such networks stand between the forecaster and his grasp of the future.

As long as it is believed that the course of population is governed by a law of population growth, the forecaster can be regarded as the neutral outside observer of the population system that is to be forecast.[109] The forecaster is external to the population system observed. The forecaster is not a part of the system and is not seen as such. That is also the way the forecaster prefers to see it. It is an idealised image of a forecaster's activity, as even Adams was aware: '*.. the personal element is by no means absent, for in the work of fitting the equation to the observed data judgment and opinion play a considerable part'* (Adams *et al.*, 1929, p. 110).

Once belief in a law is relinquished, an uneasy uncertainty with respect to the future, permitting only speculation, enters the forecasting activity. The forecaster's vision with respect to the perception of present reality and its future is blurred by the forecasters' (and society's) conceptual framework and

---

109 The following paragraphs are based on a distinction of positions made by Baumgartner (1987). Also on De Gans (1995c; 1996), who placed Baumgartner's 'positions' in a temporal perspective. From his own experience in the field of energy forecasting, Baumgartner was able to distinguish four different societal positions that apply also to population forecasting. The forecaster can either take (or be given) the position of an 'outside observer'; of an 'imparter of knowledge'; of a 'policy-maker', or of a 'competitor in modelling and implementation' (Baumgartner, 1987).

the future's uncertainty. In each calculation scheme, a parameter of unknown value has to be inserted in order to express that uncertainty (De Casseres 1926; also Chapter 7). *' .. every forecast of future fertility and mortality, and therefore also of the future population'*, so concluded Kuczynski in 1937, is *'as at best a reasoned guess'* (Kuczynski, in Honey, 1937, supplement: 'Abstract of the discussion').

When, in the Netherlands in the 1930s, population forecasts were increasingly used for planning purposes, the position of the population forecaster was initially that of the imparter of knowledge. Forecasting is seen as a scientific activity and discussion of method and assumptions is restricted to fellow forecasters, modellers, and academic researchers. Perceptions (and self perceptions) of forecasters are still of the outside observer. Further development does not of course halt at this stage. After the Second World War forecasts, in particular urban and regional forecasts, became part of policy implementation. Increasingly the actors in the system predicted have access to the model and the forecast results. This knowledge may lead the actors to adjust their behaviour, which may be contrary to the forecaster's intention. Self-fulfilling and self-denying effects may result.[110]

In a view of population dynamics where the system forecast is no longer independent of the forecast itself, social behaviour with respect to fertility, mortality and migration is thought to be under the direct and deliberate influence of the actors with structural power. The forecaster is explicitly involved in the process of translating forecast into policy, and vice versa. In the end, forecasters (modellers included), forecasting methodology, models and forecasts may become elements in a struggle for policy dominance. Modellers and forecasters, in their quest for influence and financial support,

---

[110] In fact it is inherent in Malthus' theory of population and in social Darwinism also. It was what many students of population hoped for, particularly in the 1920s in Italy (Ipsen, 1996) and in the 1930s in Germany, namely that national governments would accept their responsibilities and act accordingly. As expressed by the German statistician and forecaster Bürgdörfer, who was the director of the *Statistisches Reichsamt* in Berlin, in a reaction to the criticisms of Paulinus (see Chapter 4): *'Eine zielbewusste Staatsführung und ein im Kern noch gesundenes Volk wird sich aber mit pessimistischen Gefühlen nicht allzulange aufhalten, sondern (...) daraus den festen Entschluss ziehen, alles daran zu setzen, um die Voraussetzungen jener pessimistischen Ergebnisse zu ändern und damit auch die drohende Entwicklung ab zu wenden.'* (Bürgdörfer, 1935/1936, p. 157).

are drawn into this struggle, trying to gain dominance for their own models and forecasts by linking them to dominant policy formulation and implementation agencies.[111] A consequence of what is contained in De Casseres view of forecasting is that *'the future has become negotiable'*, as Andreas Faludi, professor of planning theory at the University of Amsterdam, said in a 1994 course on population forecasting. Or, in the words of another planner, the future is a dimension of yet unrealized possibilities; it is *'an empty container that is filled with the quasi-reality of projected images'* (Friedmann, 1973, p. 116).

It must be realised that both Faludi and Friedmann are planners and planning theoreticians, not population forecasters. To a high degree, Dutch planning owes its emergence to population and housing need forecasting. As explained in Chapter 7, in the Netherlands the population forecast was seen for a long time as the core of a general extension plan, or a regional plan. But once migration had to be taken seriously as an important factor of urban and regional development, demographic forecasting was felt to fall short as a reliable forecasting instrument. After the Second War the Dutch government became increasingly plan oriented. High hopes were put in the belief that mankind had the capacity to shape its own future. Influencing the distribution of the population over the country by influencing the size and the direction of internal migration flows, either by regional economic development policy, or housing allocation policy, became explicit government policy. Unfortunately, planners and policy makers have tended to lose sight of the demographic momentum. Such forecasters as Oly, Wiebols, 't Hooft and Methorst in the Netherlands, Cannan, Westergaard and Bowley elsewhere, and formal demographers such as Lotka, share the merit of focussing attention on the

---

[111] De Gans (1995c) makes the following temporal differentiation of the position of the forecaster in Dutch society: Until 1931 the forecaster is seen as striving for the unattainable ideal of the neutral outside observer. From 1931 to about 1966 the forecaster is seen and sees himself as the provider of knowledge in the first place. In the period 1966-1990 he is increasingly seen as a professional involved in the preparation of policy making, perhaps with the exception of national population forecasters. From 1990 onwards the process of acting as a competitor in the forecasting market comes to the fore. The time perspective in question follows closely a similar temporal ordering of the use of demographic research by Ter Heide (1992). According to Ter Heide, demographic research was considered as a *datum* for policy making with respect to physical order in the Netherlands between 1920 and World War II. From the end of the War until 1966 population was mainly seen as a problem, and from 1966 onwards population was an object for policy making.

significance of the age-sex structure by stressing its importance for future numbers of vital events and the implication for future population growth.

8.2 | Images of the Future Blurred by Conceptions

Another effect, never completely accepted in its full consequences in population forecasting, was that the link between population forecasting and 'true' Science was slackened. No longer did a forecaster of population have to be concerned with an actual law in the form of a mathematical equation. As De Casseres put it, mathematical conceptualisation was just another conceptual network which helps mankind to come to grips with reality and future reality, but like all other conceptual networks, the image presented can only be blurred (De Casseres, 1926).

In inter-war population forecasting literature, such views as those of De Casseres are rare. An exception was the criticism made by Bouthoul (not a forecaster himself) of the coercive force of population forecasts as the most imaginative offspring of modern quantitative society. This isolated position could result from the fact that demographers and demographic forecasters are closer to statistics than to the social sciences. Sauvy describes demography as just another part of statistics; demography and demographic forecasting still bear its mark.[112] Their common statistical background blinds demographic forecasters to some of the risks they run with their forecasts.

Insofar as they are involved in long term forecasting, forecasters easily fall victim to *nominalistic fallacies* (Elias, 1982/1985; Adam, 1990). They tend to overlook society's habit of housing today's contents in yesterday's words (Le Bras, 1987). The demographic vocabulary may remain the same over time, but the contents differ. People aged 65 and over in 1900, 1950 and 2050 respectively, for instance, will all be called 'elderly'. In terms of physical and social condition, however, people belonging to this age category will be in totally different stages of ageing in these three years. Similarly, an immigrant

112 'Demography as a science of population is everywhere and nowhere. It has no natural delimitations and can therefore be extended and delimited by agreement; it is limited to the observation of three factors: Age, sex and marital state – factors that are precisely determined, whereas theoretical demography is only an offshoot of mathematics and applied demography a statistical technique.'(A. Sauvy in Girard, 1958/68, p. 70).

may be seen as a foreigner through being born outside the country. But over time the immigrant's identity as a foreigner will see a drastic change under the influence of a process of integration. In terms of social distance with respect to the autochthonous population, the immigrant may become a different person. The same holds for attitudes towards marriage and divorce (e.g. Van Poppel, 1992); for attitudes of mothers towards the production of children and maternal love (e.g. Heinsohn *et al.*, 1979) and the use of 'illegitimate' fertility in a period in which cohabitation and motherhood of unmarried women are accepted social phenomena (Kuijsten, 1994).[113]

What is true for the past is true for the future. Tomorrow's reality will give different content to today's concepts. Illusory elements are inherent in futures in which wished for future developments are consciously and purposely included, for instance the inclusion of policy goals in assumption making (De Gans, 1990).

Lastly, from the point of view of symbolic interactionism, demographic forecasters are accused of failing to account for the social dimension of *temporality*. They are involved in the study of numbers, or bodies along a dimension of time, rather than in the study of people. For instance, what is particularly social about human beings is left out of most futures of population, namely social conduct in terms of *identities*. Instead, forecasters count, measure and discuss *bodies*, and they fail the sociological mandate because it is by means of identities rather than bodies that people are located and obtain meaning in society (Maines, 1978). In an organisational context identity '.. *establishes what and where the person is in social terms*' (Maines, 1978, p. 246). They obtain meaning in a structural way, insofar as identity refers to the location of a person in a social world; in a dynamic way insofar as that location is defined in terms of ongoing processes of identification (Maines, 1978; De Gans, 1993).

From this symbolic interactionist point of departure and insofar as identity is concerned Maines may be right. Demographic forecasters, however, differ little from other social or behavioural scientists in the field of population studies, as can be concluded from a recent overview of psychological literature

---

113 Man as a symboliser is excluded from analysis and forecasting.

(Fawcett, 1991).[114] Moreover, the introduction of event history analysis in demographic research has greatly contributed to the integration of social, economic and demographic phenomena and therefore to the location of people in a social world, albeit not the world of symbolic interactionism.

On the other hand, as long as the line of criticism started with De Casseres is not elaborated by demographic forecasters and is left to others, true conceptual innovation of a forecasting system in which the elements are people instead of bodies cannot be expected.

8.3 | A Break in Time, or (Dis)Continuity of Time?

Another significant consequence of the difference between the views of Thomas Adams and De Casseres is that a complete shift of the perspective of time is involved. The consequence of Adams' view of population change is that it takes place in the sphere of 'objective time', while that of De Casseres involves 'subjective time'.[115] According to the theorists of time who adhere to the objective time concept, events are to be regarded as occurring at certain clock times in the absence of tense. The events occur primarily 'earlier than', 'simultaneously with' or 'later than' each other. Time can be conceived of objectively by differentiating between earlier and later states.

---

114 The 'body' or 'person' dilemma, which basically coincides with the dilemma of demography and demographic forecasting as part of statistics or social sciences is implicitly indicated in Willekens (1990). The central issue of his oration was the search for a philosophy of man (in Dutch: *'Mensbeeld'*) in demography. Willekens asked for a new frame of reference based on a realistic *'mensbeeld'*. Willekens considers the transition in demography from a structure approach of population change to a process approach, in particular the synthetic biographic life career approach, as the manifestation of a fundamental shift of paradigm in demography (Willekens, 1990, pp. 19-20).

115 The distinction between objective time and subjective time was made for the first time by McTaggart in his book *The nature of existence* (1927); it coincided with the publications of Adams and De Casseres. McTaggart distinguished two different ways of conceiving of time; the A series (subjective time), and the B series (objective time). In the 1960s they were renamed the A and B theories of time (Denbigh, 1981). In fact, the B theory asserts that the t coordinate of physics provides a time concept which is sufficient for all purposes, with the possible exception of time experience in psychology (Denbigh, 1981). 'A' theoreticians regard the time concept of consciousness as equally necessary for application to psychology as to the physical world (Adam, 1990).

On the other hand the theorists who adhere to the 'subjective' time view maintain that 'now', or 'the present', is a real feature of the world at each location, as is 'becoming', or 'coming into being'. Defining something as past, present, or future is a relative statement. It depends on both the observer and the surrounding relationships. Therefore, time should be conceived of subjectively by involving the observer in analysis (Denbigh, 1981).[116]

Even Adams had to admit that, in the logistic population growth approach, subjective time was creeping into the perception of present reality, past trends, and emerging futures. In this, he and De Casseres have in common that in the end it is the forecaster's judgement of observed data which determines the parameters in the mathematical equation, and therefore the precise form of the logistic curve. The forecaster cannot help breaking into the flow of time, with the forecaster's presence separating past from future developments.

Through the forecaster as intermediary, society is also breaking into time. The interests and concerns of society lie in future population development, needs and willingness to invest in the requirements that make population forecasting possible (education; data collection) and to use forecasting results. At the organisational level the break into time occurs through the ways the forecasting activity is organised: Whether individuals or institutions are responsible for population forecasting; what the degree of integration of forecasting at different geographical levels is; through the position and status of the forecasting officer within the hierarchy; through the forecasting tradition of the forecaster or the forecasting institution; through the ways information in the professional field is organised and exchanged. At the individual level, the forecaster's psychology, professional education; status, forecasting history and orientation towards the future play a part. Several authors point at people's different conceptualisations of the future resulting from differences in age and past experience (Lynch, 1972, p. 90 et seq.; Ascher, 1978, Mercure, 1983, Keilman, 1990). Differences in the duration and quality of past experience

[116] In terms of recent demographic forecasting methodology: It is the forecaster's task to identify the population system that has to be forecast, to give a description of the past behaviour of the population system, to design a model that fits the dynamics of the system and to formulate assumptions of the future behaviour of the population system and select the parameter values in the equations of the forecasting model, not to mention the tasks of sensitivity analysis, implementation and monitoring (Cruijsen and Keilman, 1984; Willekens, 1990).

affect the way in which the future is conceptualised and therefore determine the course of future events.

Once the position of the forecaster comes into play as an inherent part of the system, the forecaster's own attitude towards the future can no longer be negated, neither can that of the organisation, nor of the society, to which the forecaster belongs. According to the Canadian sociologist Mercure (1983), certain fundamental orientations of life play a significant part in a person's orientation to the future. These fundamental orientations of life are the feeling of being governed by a prevailing sense of either insecurity and external control, or of security and internal control.

The insecure, externally controlled orientation can result in attitudes in which the future is seen as either *the unavoidable*, or *the foreseeable*. The secure, internally controlled orientation can result in attitudes in which the future is seen as *the continuation* (of current conditions), *the step by step* (achievement of life career targets), or *the possible* (the achievement of targets posed). People who consider the future as *unavoidable* are characterised by a fundamental orientation of life which is directed at *preservation*. Career and life planning having no use; expectations are passive; the future just comes, it is unavoidable. People belonging to the second category are governed by a similar prevailing sense of insecurity and of being externally controlled and by a similar basic orientation of life that is directed at preservation; career and life planning are seen as of no use. Expectations and actions however are directed by what can be foreseen (the *foreseeable*).

The fundamental orientation of life of the three remaining attitudes is directed at *conquest*. The orientation derives from a prevailing sense of security and internal control. Attitudes towards the future include: Confidence that the prevailing situation will continue, which makes people's need for career and life planning superfluous. Secondly, one can adopt the attitude of career and life planning that is planned step by step. Lastly, life can be seen as a systematic conquest of the future by means of systematic career and life planning. The life and career targets posed from the beginning are pursued relentlessly, because the future will see its fulfilment (the possible).

Of course, representatives of each category in its purest form would be hard to find. Moreover, people may find themselves in different categories in different stages of their lives. Forecasters can be assumed to be found in all these categories and there may be some relationship between orientation towards the future and assumption making, a relationship that becomes more pronounced with the extension of the horizon of the forecast. The longer the forecasting period, the more dependent becomes assumption making of a theory of future population and the more can hopes and fears blur the realistic quality of expectations (Friedmann, 1973, p. 116). Although it risks falling into the trap of *ecological fallacy*, Mercure's typology might also apply to forecasting organisations and societies in various periods in history.

Nowadays, responsibility for most national demographic population forecasts rests on bureaucratic, hierarchically structured organisations. Some sort of relationship between the position of the forecaster within the organisation, for instance regarding the degree of freedom in taking risks in assumption making, and future orientation of the organisation, cannot be excluded. The accuracy of a forecast has also to be evaluated in terms of the 'assumption making behaviour' of individual forecasters (Ascher,1978; Keilman, 1990). Ascher suggested studying the sociology and psychology of forecasters. Institutional affiliations, training and the climate of opinion in which forecasters work should all be seen as potential sources of explanation why some forecasts are more accurate than others. Along a similar line of reasoning, Keilman suggested investigating how individual forecasters select trajectories for key demographic indicators and the extent to which their decisions are based on demographic and non demographic information; how they are influenced by the opinions of others; what the impact of the forecaster's own demographic situation and history is on the decisions; personal reactions to major errors in earlier forecasts in which the forecaster was involved.

Forecasters who understand that they are acting in subjective time have three options should they wish not to renounce forecasting. They could limit themselves to simulations of alternative futures on the basis of the demographic models available. Or, they could search for new models of social reality in which demographic change has its proper place. Or, they could invest in what Karl Popper called *technological predictions* instead of *historical prophesies* (Popper, 1957/1966, p. 44; Van Seventer and Vossen, 1997; Le Bras, 1987). Popper criticised forecasters for being too concerned with accurate predictions and further sophistication of forecasting methodology in social sciences (which he called historical prophesying), while they ought to be directing themselves to the determination of the kind of actions needed in order either to avoid the consequences of change, or to prepare for such consequences. In the case of demographic forecasting such an approach to the future would imply the mobilisation of the available knowledge of the inertia of demographic process and concentration on the assessment of the upper and lower limits of future population. Forecasters could learn from the engineers who design Dutch sea and river dikes. Their constructions are not based on predictions of the exact date of a storm tide endangering the low lands of the Netherlands, but on calculations of the strength and height of the dikes needed to prevent flooding in the next three hundred or thousand years.

8.4 | George Herbert Mead's Philosophy of the Present

That the transition from 'law' to 'reasoned guess' involved a revolutionary shift in the founding concept of forecasting, as indicated by the substitution of objective time by subjective time, should now be clear. Instead of a mere 'earlier than' and 'later than', the predicting person and the predictor's present became central in the forecasting process as did, by implication, the predictor's past and attitude towards the future. In the preceding chapters, implicit and explicit examples of the impact of history on population forecasting abound; in terms of the past experiences of the forecaster, the forecasting organisation and the society to which forecaster and organisation belong, consideration is also asked for the basic temporal concepts entering population forecasting as a consequence of the shift. Up to now little attention has been paid to this subject in demographic forecasting literature.

Demographic forecasters belong to the group of specialists in time or *'along a dimension of time'* (Giddens, 1984, p. 354). Time is the all important dimension in forecasting, pervading demographic analyses and forecasts in the guise of period time, to be compared with physical or natural time in the physical sciences (symbol 't') or in the form of duration, or age (symbol 'x'). Demographers and demographic forecasters are convinced of the necessity of anchoring the lifetime of birth cohorts with calendar years and dates in the history of total population, because they know that the events occurring in these birth cohorts –and by implication in total population in a specific period of time– result from the joint interaction of age, period, and cohort influences. In cohort analysis, they have taken the lead; they are among the first to be familiar with the cohort as a necessary concept in the study of social change (Ryder, 1965; Becker, 1992). They even have knowledge of eternity, because they are familiar with the process of demographic metabolism, the ongoing replacement of vanishing cohorts by new ones with demographic populations renewing themselves and striving for eternity (Ryder, 1965) or, in the case of insufficient replacement, to fears of extinction and race suicide (Kuczynski, 1928; Soloway, 1990).

The basic concepts of subjective time rarely play an explicit part in demographic studies. Although, more than most social scientists, demographic forecasters are specialists *in* time, for this very reason they are not specialists *of* time.[117]

A contribution to a more fundamental consideration of the basic temporal notions (present, past and future) in terms of subjective time theory is provided by George Herbert Mead (1863-1931).[118] The core element of Mead's philosophy of the present is the axiom that the present is the only locus of reality. Reality cannot exist except in the present (Mead, 1932/1980). Neither past nor future have existence in or of themselves (Mead, 1929; Maines *et al.*, 1983; Joas, 1980/1985; Pronovost, 1989).[119] Mead's philosophy of time is therefore fundamentally different from philosophies of the past or the future, which tend to regard the present and the future as mere concatenations of effects resulting from causes which are effective now and for all time (Joas, 1985).[120] The distinction of the present from the past and the future is funda-

---

117 Times are changing, however. In 1993, a conference on the subject of time and demography was organised by the *Chaire Quetelet* of Louvain la Neuve in Belgium (see Vilquin, 1994).

118 Interest in Mead's contribution to a social theory of time has grown in recent years, as is apparent from recent overviews of literature on the problem of time in sociology and sociological theory (Maines *et al.*, 1983; Pronovost, 1989; Adam, 1990; Bergmann, 1992; Novotny, 1992). Sections 7.4 and 7.5 of the present chapter are mainly based on Mead (1929; 1932/1980) and a few recent studies of Mead, particularly Maines *et al.*, (1983), Joas (1985) and Adam (1990).

119 For a discussion of the question whether the present exists only in human experience as awareness of the present, or objectively in the world-at-large, inanimate as well as animate, see Denbigh (1981, pp. 51-68).

120 According to Joas (1985, p. 171), Mead originally considered the nature of the present as a now point, without temporal extension. Empirical psychology convinced him that a definite minimal temporal extension is required for perception and that therefore the present without a time dimension could only be an idealisation of this psychical present. For that reason, Mead started to use the term *specious present* instead. In Anglo-Saxon literature this term had gained acceptance as an indication of this minimal temporal extension. An example of a contemporary clarification of the notion 'specious present' by C.D. Broad, in 1923, runs as follows. If one takes sensing to involve a process with

mental. In the words of Mead: If the present were spread so that it covered more events, '.. *taking in some of the past and conceivably some of the future..*', the events so included, would belong not to the past and the future, but to the present (Mead, 1929).

Another essential element of Mead's philosophy is the notion of the objective existence of *novelty*, embodied in the concept of *emergence* or *emergent event*. The present becomes manifest as present by an emerging event. The emerging event is at the origin of all structuring of time. Without events, there would be just 'passage' and time would not be perceived.

Although an emergent event is located in an unbroken causal chain lying in the past, because of its novelty its emergence cannot be deduced from the causal chain before it emerges; the emergent event is conditioned, but not determined, by the causal chain. After the event has emerged ".. *it compels 'the reconstructive interpretation' of this causal chain*" (Joas, 1985, pp. 176-177), which means that the event's causal history has to be rewritten in such a way that the new event fits into an unbroken causal chain. The very quality of the novelty of the emerging event implies the existence of a temporal and causal discontinuity. Although continuity forms the background for the novelty (Mead, 1929), an emerging event is a break *in* this continuity and at the same time it is a revelation *of* the continuity. Without this break in the continuity of time, successions of events and continuity could not be experienced. Everything that emerges has continuity, but *only after it has arisen*.

An essential aspect of the notion of continuity as a succession of events is that it can only be established by operation of the mind, by acting persons who recognise continuity as a succession and make it intelligible as such (Maines *et al.*, 1983). The past must be reconstituted for there to be continuity. This act of reconstitution of the past connects successive presents. In other words: That which was unconnected becomes connected by the merging of successive presents.[121] In the process of connecting what was unconnected, the quality

a finite duration, then anything an observer can sense at any moment '.. *stretches a little way back behind that moment*' (Broad, in Krausz, 1986, p. 49). It is not possible to draw an absolutely sharp line between sensing and remembering.

121 It would be a mistake to conclude that Mead denied the objectivity of time or causality. An emerging event is located in an unbroken chain lying in the past. But because of its novelty it could not be deduced from this past before it had emerged. It compels a reconstructive interpretation in order to be fitted into this causal chain (Joas, 1985).

of novelty in the emerging event in the present, which was the very essence of it, is lost forever and can never be brought back: *'The primal break of novelty in passage is gone with the past present and the problem of bridging the contingent factors is before the mind'* (Mead, 1929, p. 240).

The past itself should not be mistaken for a past present: The past is not a present after it has merged into another present. The moment a present merges into another present, the past present becomes irrevocable. Events cannot be undone, thoughts cannot be un-thought and knowledge cannot be undone. They all become irrevocable. The moment continuity is re-established by fitting it into a causal succession of events, its novelty is lost and the past present becomes unknowable.

The following example may clarify this. The majority of Dutch demographers and demographic forecasters of the mid 1960s were taken by surprise by the rapid fall in fertility from the second part of that decade onwards. The issues of the national population debate at that time were of a different nature. First, there was the issue of the differential fertility between the Roman Catholic part of the population (characterised by a relatively high fertility) and the rest of the nation and the consequences for the future composition of the population. Next, their was the issue of the increase in the size of the population (estimated at about 20-21 million inhabitants in the year 2000) and its consequences for housing and physical planning. This issue was put on the agenda by the planner Prof. Jacq. P. Thijsse at the Annual Meeting of 1963 of the Dutch Demographic Society, and a few years later by NCBS (Thijsse, 1963; Van den Brekel, 1996). Only after the fall in fertility rates in the second part of the 1960s did it become clear that indications of the imminent fertility decrease could have been perceived much earlier.[122]

---

122 Recent history provides other even more clarifying examples, although not from the field of demography. Up until the end of the year 1989 historians, social scientists and politicians proved convincingly by force of arguments based on plausible analyses that the partition of Germany would continue for several more decades. Now, after the turbulent events of the collapse of the German Democratic Republic and since the unification of the two Germanies, a causation (in the form of all kinds of explanations) for this unexpected and until recently unimagined course of events is being sought. Within a short time explanations will emerge in which the dissolution of the German Democratic Republic and the changes in Eastern Europe and the Soviet Union will be presented with retroactive effect as unavoidable and clearly predictable. In discussions resulting from recent events in Middle and Eastern Europe, including what is happening in former Yugoslavia, it is assumed that in future international conflicts will no longer

8.5 | Dimensions and Forecasting Implications of the Past in Meadian Philosophy

Mead's philosophy of the present has direct and unexpected consequences for the perception of the past and, by implication, of the future. At least four different dimensions of the past can be distinguished in his philosophy: The social structural past, the implied objective past, the symbolically reconstructed past and the mythical past (Maines, Katovich and Sugrue, 1983). Although for analytical reasons the dimensions are discussed separately, there is in fact a constant interaction between them.

The *social structural past* is the past with which demographic forecasters are most familiar. It is the past generally implied in demographic analyses and forecasts. The social structural past structures and conditions the events that emerge in the present. In the words of Mead: *'That which is novel can emerge, but conditions of the emergence are there. It is this conditioning which is the qualitative character of the past as distinguished from mere passage.'* (Mead, 1929, p. 236). Because of this dimension, causal relationships can be established between successive events, and thereby distinctions can be made between continuity and causality processes. Social structural pasts only establish probabilities for what will take place, but they do not determine what is taking place (Maines *et al.*, 1983).

The *implied objective past* is the past that can be reconstructed from the facts as experienced in the present. The implied objective past is the past that logically *must* have been in order to be present in experience as a past (Mead, 1929). Out of the past are selected only those events which could have occurred, given the present structure or arrangement of events. We believe in the reality of past events, because they connect in a credible way with the reality structure of the present. For instance, one must have been born and been married to be divorced in this year. Without this dimension of the past, history writing would not be possible.

---

take place along partition lines based on ideologies, but along partition lines based on cultural differences, whether perceived as such or wilfully created. From the reconstructed past and the re-established continuity, a new future is reconstructed. For the time being, one concept of reality, namely the belief that future international society will still be in need of concepts based on 'enemy images' awaits disturbance by an emerging event.

Imagery belongs to the present, but there are certain sorts of images which belong to our pasts. We are confident that these images are not just fantasies that belong to our present, but belong to the past *'..because they fit in (....). The assurances which come from a remembered occurrence come from the structures with which they accord'* (Mead, 1929, p. 237). They are credible, for they fit present reality as we perceive it.

The *symbolically reconstructed past* is the dimension most often overlooked in demographic forecasts based on long term trend analyses. The neglect of this dimension of the past is the source of most nominalistic fallacies occurring in long term historical and trend analyses. For this reason, and the implications for scenarios of the future, this dimension of the past is further elaborated. The symbolically reconstructed past involves a redefinition of the meaning of past events in such a way that they have meaning in and utility for present reality. The symbolically reconstructed past is a backward overflow from the present, which at the same time modifies the view of the future. In the dimensions of the past discussed the present results from the past, while in this dimension the present makes the past possible. Present reality, which can only be experienced as present reality by the novel character of the emerged event, forces the reconstruction of the past and the establishment of a new continuity (and therefore a new past) in such a way that the emerged event finds its 'natural' (causal) place. From this newly established causality new forward continuities, adjustments, plans and actions for the future result.

In Maines' interpretation of Mead's philosophy, the present is not a watershed between past and future, but the location of reality from which pasts and futures emanate. At the same time, the emergence of new events that create (new) present reality is conditioned (not determined) by these established continuities. The past is not what has happened, but what is still happening. The *what it was* is established through the *what it is*. Each present therefore must reconstruct its past, and there are as many reconstructed pasts as there are novel events (Maines *et al.*, 1983).

Similar notions can be found independently of Meadian philosophy. Archaeologists, historians, restorers of ancient monuments, landscape geographers and many others know that light in the present is only shed on the past by the presence of memory, tradition, history and material relics of earlier times. However, the past that is revealed is not simply what happened; it is in large measure a past of our own creation, moulded by selective erosion, oblivion, and invention. ".. *we cannot avoid remaking our heritage, for every act of recognition alters what survives. We can use the past fruitfully only when we realize that to inherit is also to transform.*" (Lowenthal, 1985, p. 412).[123] The management of change and active use of (historic) remains for present and future purposes are to be preferred to an inflexible reverence for a sacrosanct past, says Lynch (1972, p. 64): *'The past must be chosen and changed, made in the present. Choosing a past helps us to construct a future.'*.

Similarly, the historian Lorenz, states in his investigation of the relationship between critical or analytical philosophy of history and historiography, that the material object of historical science is no longer there; it cannot play a part in the selection of explanations. Only *vestiges* survive in the historian's present. The historian needs hypotheses to be able to make sense out of these vestiges (the historical facts). These hypotheses are closely related to the social and societal context within which they are generated (Lorenz, 1987).[124]

A demographic forecaster living in a specific period of time operates within the framework of a variety of continuities, including those relating to the existing forecasting tradition, as a set of psychologically and socially conditioned ways of looking at reality. These continuities make it possible to set goals and plans for the future. At the same time, the forecaster has to deal with the novelty of the emergent events of present reality. This may result in finding new concepts of looking at past and present reality; in finding new facts; or in the selection, ordering and interpretation of already existing facts in a different way. In doing so not only are new *system identifications* of present reality being made, but also new pasts and futures resulting from it.

The concept of the *second demographic transition* can be taken as an example (Lesthaeghe and Van de Kaa, 1986; Van de Kaa, 1987). According to this concept, in most Western industrialised societies a transition from an altruisti-

---

123 A 'past present' is a past that is closed to influences. In the terminology of Lowenthal it would be called a 'dead past'. The symbolically reconstructed past of Mead's philosophy would be a 'living' past in the terminology of Lowenthal.

124 Although in Meadian philosophy thought processes are essential for the reconstruction of the past, change is not just something existing only in the mind. Changes are going on in the universe, and as a consequence of these changes the universe is becoming a different universe. Intelligence is just one aspect of this change. It is a change that is part of an ongoing living process that tends to maintain itself. But the mind is only able to establish causality and continuity (Mead, in Maines *et al.*, 1983, p. 163).

cally driven attitude towards children into an individualistically driven attitude has taken place since the 1960s, resulting in a net reproduction rate below replacement level. Expectations of future developments of fertility should then be considered within the context of expectations of the future course of the process of individualisation. If, however, larger historical and even evolutionary biological processes in the contents of such concepts as altruism and egoism were considered (Cliquet, 1991), or if the theory were not limited to cultural changes since the 1960s (e.g. Kronjee, 1992), the context of the future course would have been different, as would the expectations. According to Kronjee, cultural changes since the 1960s are useless as an explanation of the observed decrease in average number of children per generation over a period of a hundred years. The decrease should be explained in terms of a growing concern for care and education within the family. A rise in the burdening of tasks per child is compensated by a decrease in the number of children per family (Kronjee, op. cit.). If Kronjee is right, future fertility expectations should be considered within the framework of future developments of the burdening of parents by tasks in the field of child care and education. Moreover, the basic idea of a transition from an altruistically driven attitude towards children into an individualistically driven attitude already existed in Falhlbeck's characterisation of demographic reality at the beginning of the 20th century. Adherents of the second demographic transition theory consider the second demographic revolution of the 1960s and 1970s responsible for the present below replacement level situation of population in many West European countries. This, however, coincides with one of Fahlbeck's possible futures, namely negative population growth resulting from his expectation of the future course of the vital rates of population growth (Fahlbeck, 1905; also Chapter 3).

Lastly, the dimension of *the mythical past* is only indirectly implied in Mead's philosophy. A mythical past is a fictitious past; it is not a re-creation of a past present, but a wilful new creation that is not empirically grounded. Mythical pasts are present-centred creations, created with the purpose of controlling and shaping behaviour towards future lines of action. They provide a manipulative dimension to social relationships, because they are created for manipulative purposes. Mythical pasts represent *pseudo-objectivity*, used for the benefit of only certain actors at the expense of others. This past is different from an implied objective past. Instead of being regulated by the force of credibility, however, a mythical past is governed by the force of manipulation and is based on 'make believe' (Maines *et al.*, 1983).

In demographic forecasting, however, it is difficult to draw clear cut lines between wilful creation of pasts that are not empirically grounded and conscious or semi-conscious distortion. Such distortion can hardly be avoided. In trend analyses, for instance one has to decide whether incidental changes in frequencies should be taken into account or eliminated.[125]

All kinds of conscious or semiconscious manipulations could be categorised under the concept of mythical past; assumption drags, selection of explanatory theories, the selection of analytical and forecasting instruments, the use of seductive metaphors, analogies, precedents or concepts in which essential elements of the system to be forecast have not been accounted for, as in 't Hooft's stubborn clinging to his metaphor of the conveyor belt (Chapter 5). Moreover, mythical pasts do not lead automatically to mythical futures. The very essence of a future is that it has a mythical quality. The future cannot escape conditioning, not to be mistaken for determination, from the present and the past belonging to that present.

Given the conditioning, the future is open to influence by present action, whether generated by the present perception of implied objective pasts, or of mythical pasts. Demographic forecasters taking their job seriously will take into account the goals for the future policy makers set. In short term and middle term goals, arguments from trend analysis and preliminary planning research play a dominant part. In long term goals, ideological visions may take over, as may even utopian goals as the horizon widens (Friedmann, 1973). Setting goals implies manipulation of the future until it accords with present day needs, desires and visions. But even visions are conditioned by the past; they are not completely disconnected from it.

The 'old' paradigm of the study of future populations allowed forecasters to believe that future population growth was governed by a law of population growth and that the future course of population could be predicted with confidence. Under the old paradigm, the forecaster could in principle appeal to the authority of 'laws' regulating future population growth. The best that forecasters could hope for under the new paradigm was to make a reasonable guess, based on an improved insight in the determining factors in the processes

---

125 An example from recent Dutch history is the peaks in the curve of the number of immigrants from Surinam to The Netherlands that resulted from changes in the rules of admittance.

of demographic change. Demographic forecasters knew they were bereft of such an authority. What remained was the plausibility of their arguments by force of their professionalism and existing population theory. No longer was the forecaster the mediator who reveals how the law of population growth works in a specific situation. Under the new paradigm, the presence of the forecaster in the specific period of time in which the forecast is made, the associated perceptions and conceptions of the present reality, give form to the specific past and future that emanate from the forecasters specific present.

While the perspective of time of the old paradigm was the 'earlier than' and 'later than' of so-called objective or physical time, the time perspective of the new paradigm is that of subjective time, characterised by past, present and future. In this perspective of time, the forecaster takes a central position as the person who establishes continuities from the past into the future. It is the merit of Mead's philosophy of the present that it gives a specification of the central position of the forecaster in the forecasting process without neglecting the objective aspect of time as a part of the complex interplay of dimensions of time the forecaster operates in.

9. Conclusions

9.1 | Introduction

The transition to modernity of population forecasting, from the first emergence of demographic forecasting in 1895 with Cannan's forecast of the population of England and Wales and it becoming the new standard method of national population forecasting, took about five decades. To a large degree the new methodology owes its existence to the wish to improve the quality of the debate on the population problem by giving it an objective scientific foundation of the ongoing debates. Only much later it became clear that the new methodology could also be used for decision making purposes.

Within a decade after its (re)emergence in the Netherlands and other European countries in the 1920s demographic forecasting found its first application in urban and regional planning practice. Town planners in Amsterdam and Rotterdam had discovered that Wiebols' pioneer forecast of 1925 could be used in urban and regional settings by inserting internal migration in the calculation schemes. However, they prudently refrained from following the elaborate directives for a further sophistication –based on a rigid, demographic component specific occurrence/exposure approach– which were indicated in Wiebols' PhD thesis and elaborated in his letter to the statistician Van Zanten.

By 1945, demographic forecasting had become the new standard method for the calculation of the future size and structure of the population in the Netherlands. The method had substituted Malthusian geometrical forecasting of total population. The calculation of the doubling time of a population with geometrical growth methodology, which had been the main concern of 19th and early 20th century economists, had been replaced by a method that was better equipped to serve the planning minded post-World War II era.

9.2 | Dutch Pioneers of Demographic Forecasting

Our history started with the assertion that the Dutch pioneers of demographic forecasting in the inter-war period, Oly, Wiebols and 't Hooft in the 1920s and Van Lohuizen, Delfgaauw and Angenot in the 1930s were international pioneers as well, standing comparison with their international contemporaries.

The assertion found support in a statement made by Angenot in the early 1950s that the domestic (Dutch) forecasts were much better than the foreign ones (Chapter 5). And in 1960 Godefroy held the opinion that Wiebols was the first in the world to make a cohort-survival forecast (Chapter 1). These confirmations of the premise of an inter-war Dutch pioneer position in the innovation of population forecasting methodology have proven to be right. That is, if the Dutch endeavours are compared with other contemporary forecasts; or, in comparison with earlier foreign ones from the point of view of methodological sophistication and elaboration. For that reason it can be concluded that the names and deeds of the Dutch pioneers are unjustifiably absent from intellectual histories.

However, if the criterium 'novelty of contribution' is taken as point of departure, the pioneering endeavours of Cannan, Fahlbeck and Westergaard went before them. The arguments supporting these conclusions are presented below. A summary of the nature of the contributions to the modernisation of population forecasting made by the national and international pioneers of demographic forecasting referred to in the preceding chapters, listed by name and year, is presented in *Table 9.1*.

The international history of demographic forecasting begins with Cannan's quantitative prediction of the future growth of the population of England and Wales in 1895; Fahlbeck's trend extrapolation of the vital components of population growth of 1905; Westergaard's horoscope of the population in the 20th century of 1907. Together, these contributions heralded the emergence of the new demographic approach to population forecasting.

In terms of novelty of method no later forecast can stand comparison with that of the English economist Cannan, although as early as the 18th century the Dutch actuary Kersseboom had already indicated how to proceed.Cannan's forecast heralds the true beginning of the era of modern population forecasting.

All the ingredients of cohort component methodology are there, albeit in a rudimentary form.

Cannan's long term forecast was intended as a direct attack on current official geometrical forecasting methodology. He demonstrated that it was possible to predict with a large degree of likelihood what the direction of future population growth would be if the future influence of the past and present age-cohort structure were taken into account.

However, no early nor contemporary forecast can stand comparison with that of the Dutch economist-statistician Wiebols, in 1925, in terms of the demographic specification, suitability for modelling purposes, allowance for integration of migration, and the demographic analytical foundations of its assumptions. Not even that of the English statistician Bowley of 1924, whose forecast is generally taken as the first modern demographic forecast of the interwar period in intellectual histories. Wiebols was the first to make a true cohort survival forecast and to specify the lines along which national and even urban cohort-component forecasting, based on the principle of age-sex specific sets of occurrence/exposure rates, were to proceed.

Wiebols' forecast was not the only Dutch demographic forecast of the early 1920s. The calculations of the actuary Oly of 1924 also bear comparison with those of Bowley. Both Oly and Bowley applied survival probabilities from recent life tables and stable population theory; both started from the inter relationships of the future numbers of such demographic events as births and deaths and the age-sex structure of the population. Oly stands out because of the specification of his calculations. In contrast with Bowley, Oly gave a specification of the range of the future size of the population of the Netherlands by calculating two different futures. One was theoretical and one more realistic. The population resulting from a continuation of the observed decrease of the birth rate (and constancy of the survival rates) constituted the practical exercise. As a theoretical exercise he calculated how long the general fertility rate could continue to fall before the conditions of a stationary population were to occur.

The international pioneering position of Dutch forecasters did not halt in the 1920s; that position was maintained in the next decade. Van Lohuizen and Delfgaauw's (1932) municipal forecast of the population of Amsterdam, and

Table 9.1. The pioneers of demographic forecasting, the year of their forecast and the innovations they made, 1895-1945 (the names of the Dutch forecasters are in italics)

Name	Year and innovation(s) / assumptions
E. CANNAN (1861-1935) England	1895 *national forecast* first (rudimentary) demographic forecast * cohort approach * 'survival-in-England-and-Wales' proportions (combined mortality and migration approach) * future number of births constant * age-period-cohort diagram
P. FAHLBECK (1850-1923) Sweden	1905 *national forecast* * thorough assumption making continuation fertility decline * extrapolation of components of natural growth (birth rates and death rates) * open-ended final situation
H. WESTERGAARD (1853-1936) Denmark	1907 *qualitative 'scenario' for Europe and the world* * thorough assumption making continuation fertility decline * thorough analysis of consequences future course of mortality and fertility for: 1) future age structure and its implications; 2) differences in stage of demographic transition of various nations; 3) resulting impact on size and direction of trans-continental migration flows; 4) internal migration (sub-urbanisation)
Joh.C. OLY	1924 *national forecast* * stable population theory * application of general fertility rates in assumption making (though not in calculation) * two variants: 1) constant number of births; constant survival rates. 2) constant survival rates; dynamic (decreasing) birth rates
A.L. BOWLEY (1869-1957) England	1924 *national forecast* * stable (stationary) population theory * constant birth rates * constant survival rates

G.A.H. WIEBOLS *(1895-1960)*	1925 *national forecast* * age/sex structure * general fertility rate * constant cohort survival rates * presents calculation of net reproduction rate (NRR); application of NRR in assumption making * advocates use of age specific fertility rates * presents elaborated (though theoretical) cohort-component model for national and urban forecasting on occurrence/ exposure basis
F.W. 't HOOFT *(1896-1941)*	1927 *national forecast* * longitudinal cohort survival approach * 'tidal wave' concept
G.Th.J. DELFGAAUW *(1905-1984) and* *Th.K. Van LOHUIZEN* *(1890-1956)*	1932 *urban population forecast (Amsterdam)* * general fertility rate * age specific migration in 'cohort survival'-model * application of age specific head-ship rates
L.H.J. ANGENOT *(1901-1979)*	1934 *national and urban population forecast (Rotterdam)* * age specific fertility rates * age specific migration (in Rotterdam forecast) * introduction of mathematical matrix approach in demographic forecasting

Angenot's (1934) forecast of the population of the Netherlands and of the Rotterdam Harbour Area also belong to the international vanguard. They included migration in cohort-survival forecasting methodology and demonstrated the utility of the methodology for housing need calculations with age specific headship rates (in the Amsterdam forecast). Angenot was not only the first to use age specific fertility rates in his calculation schemes but also the first to make an attempt to formal matrix modelling.

Pioneers of demographic transition theory

That the early international forecasters (Cannan, Westergaard, Fahlbeck) spearhead the intellectual history of demographic transition is an unexpected but interesting by-conclusion of the comparative analysis of early attempts to demographic forecasting as well. This is particularly true of Westergaard's 'horoscope' of the future course of Europe's population. Westergaard pictured in an imaginative way the demographic transition to come: The effect of the fertility decline on the future age structure of national populations and its socio-economic implications, and the effect of differences in the pace of fertility decline between nations on international migration flows. Fahlbeck's long-term extrapolation of the observed time series of the rates of the vital components of demographic growth stands out because of his refusal to start from the presumption of later theorists of demographic transition that, eventually, population would arrive at some stationary stage.

9.3 | The Fear of Speculation

It is argued in Chapter 1 that the transition to modernity of the study of future population took on the guise of a change of standard forecasting methodology. It is presumed that the transition might have involved a major change of fundamental belief with respect to the nature of population growth. The belief that population growth is determined by scientific law was substituted by the acceptance of the fact that the future of population growth is open to speculation and that at best the direction of future population growth can be indicated: Based on knowledge of the interaction of population structure and fertility and mortality rates (for the short and middle term) and theories of population (for the long term). The 'certainty' of geometrical, Malthusian extrapolation was substituted by speculation on the basis of demographic forecasting.

The difficulty to accept speculation as a serious statistician's activity would account for the fact that statisticians and economists took so long to become interested in demographic forecasting, notwithstanding the advantages of the approach in understanding future population dynamics were so obvious. Demographic forecasting emerged in 1895 and thrust forward in national population forecasting a quarter of a century later, in the 1920s. From the mid 1920s onwards, there was an outburst of national population forecasts in several countries of Europe based on a demographic analytical approach.

There is no doubt that, at the turn of the century, most statisticians and economists with a statistical background were familiar with the significance of demographic analysis in understanding population dynamics. Demographic analysis was developing fast. Such outstanding statisticians as Böckh, Farr, Lexis, Verhey, and Körösi contributed to its development. The continuous decrease of mortality and the observed trend bend of fertility made the assumption of continued constant growth rates highly questionable, thereby undermining population forecasting based on geometrical growth methodology. But departing from a belief in a law of population was a different matter.

There are several indications of resistance to renouncing belief in population laws among economists, statisticians and actuaries alike. Signs are visible from the very moment of the emergence of demographic forecasting. Both Cannan and Westergaard referred to the difficulty the scientific statistical and actuarial establishment had in renouncing beliefs in laws of population. Cannan and others had to campaign for some time against mainstream geometric forecasting methodology as practised by the Registrar General's Office. In France forecasting was completely abandoned until the late 1920s. The ambivalence with respect to demographic predictions at that time is clearly reflected in the way Westergaard's 'horoscope' was discussed by the Dutch statistician Verrijn Stuart in 1910. Verrijn Stuart was impressed by its imaginative qualities, but he disregarded it as a serious activity because of its speculative character.

The ambivalence is reflected in the views of the pioneers themselves. They were firmly convinced of the added value of their demographic analytic approach in terms of understanding population dynamics, but at the same time they were overzealous in emphasising the speculative quality of their attempts. Westergaard's use of the term 'horoscope' speaks for itself. In fact, accentuating the speculative character of demographic forecasting is a common factor in all the innovative forecasting efforts of the first decades of the 20th century. In debates on the method of forecasting the future course of population in the 1920s, both Yule, mathematical statistician and proponent of the logistic growth methodology (the law of logistic population growth), and his opponent Bowley, the proponent of demographic forecasting, stressed the uncertainty of predictions of the future course of population.

Because of its speculative character, many statisticians did not consider demographic forecasting to be a serious part of statistics. This is easy to understand. The statistical establishment, that of the Netherlands in particular, consisted mainly of the chiefs of statistical institutes who were responsible for public statistics and their reliability. Accepting demographic forecasting as part of statistics would imply a change of paradigm.

However, not all statisticians had such a restricted view of statistics. An exception must be made for the representatives of the mathematical school of statistics, who were familiar with probability calculus. The innovation of forecasting methodology in the 1920s came from the direction of mathematical statistics. Holwerda, who was the supervisor and the source of inspiration of Wiebols, in the Netherlands and Bowley in England were adherents of the mathematical school.

9.4 | The Absence in History of Dutch Innovative Endeavours Explained

Remains to be explained why, firstly, the pioneering endeavours of Dutch population forecasters of the 1920s and 1930s were ignored in contemporary international literature; secondly, why they are absent from recent histories compiled in the Netherlands and elsewhere.

The explanation of the latter of these questions is the orientation of historians to overviews of the inter-war period, wherein the performances of Dutch forecasters are absent. As far as recent Dutch literature is concerned, the absence can presumably best be explained from the one-sided orientation upon Anglo-Saxon formal mathematical literature.

The explanation of the absence in contemporary overviews, however, derives to a high degree from the position taken by Dutch participants at international meetings where the debate on the issue of the future course of population took place. International acquaintance with the achievements of the Dutch pioneers depended on the actions of these Dutch internationals. Methorst, Verrijn Stuart and Van Zanten were in the best position to further such recognition. Considering the international and national position of these men, one would have expected them to be eager to report their compatriots' achievements in population forecasting in international circles. Why did they refrain from doing so?

The Dutch 'internationals'

The community of Dutch statisticians and economists interested in the study of population in the first four decades of this century could consider itself privileged in having C.A. Verrijn Stuart, H.W. Methorst and J.H. van Zanten as representatives at international scientific occasions where the issue of the future course of population and its consequences was discussed. Particularly significant were the occasions provided by the International Statistical Institute (ISI) and the International Union for the Scientific Investigation of Population

Problems (IUSIPP; established in 1928 after the first World Population Conference of 1927).

Though few in number, the influence of these men should not be underestimated. Verrijn Stuart and Methorst held key positions in the ISI, Methorst also in the IUSIPP. As the head of the Amsterdam Bureau of Statistics, Van Zanten was in close contact with colleagues in similar offices in other countries. Moreover, when they perceived them, these men had a keen eye for the interests of their country and where necessary they would combine forces to safeguard Dutch concerns. Verrijn Stuart and Methorst succeeded, for instance, in having the seat of the Permanent Office of ISI located in The Hague in 1913. Whenever continuation of the location was challenged, their joint action was able to counter the threat.

The Dutch members of the international statistical establishment had achieved status, rank and position as directors of national or municipal statistical offices. The involvement of these men in statistics was primarily to safeguard the reliability of public statistics. In particular –and most outspoken– Verrijn Stuart and Van Zanten belonged to schools of statistics that largely denied mathematics, in particular the calculus of probabilities, a place. They shared the opinion that statistical probabilities and speculation about the future had little in common with statistics. Statistics had to do with description, not with probabilities.

Renouncing belief in the laws of population (law of mortality; law of population growth) and consequently accepting uncertainty with respect to the future course of population implied for these men that the speculation of the future direction of population inherent in demographic forecasting methodology did not belong to the domain of statistics. Allowing demographic forecasting to become the statistician's task would endanger the nation's trust in the reliability of its statistics. Should one be interested in the future direction of population, the Net Reproduction Rate was a good enough indicator, thought Van Zanten in the late 1930s.

In Verrijn Stuart's opinion predictions, as for instance Westergaard's horoscope of the future of Europe's population (1908), should be treated with suspicion because they were mere speculations. Verrijn Stuart was not opposed to demographic forecasting, but he lacked the interest in its methodology and it is doubtful whether he ever thought demographic forecasting to be taken seriously. His view of statistics prevented him from acting as a propagator of knowledge of his compatriots' contributions to forecasting methodology. Studies of future population interested him only insofar as they confirmed his Malthusian conviction that the future growth of the population of the Netherlands would come to an end, because otherwise the nation would fall victim to impoverishment and other disasters. The calculations he himself made of the doubling time of the population with geometrical growth methodology were never intended as forecasts, but only as warnings that population growth could not continue at the pace of the recent past.

Van Zanten was rigidly opposed to population forecasting by statisticians and statistical offices and he never changed his mind. He therefore had no reason whatsoever to propagate the achievements of Dutch demographic forecasters in international circles. At the ISI conference of Tokyo in 1931, instead of furthering international knowledge of Dutch achievements, he used his position in an effort to turn the tide when he discovered that the mainstream attitude among statisticians had become more favourable to forecasting by statisticians and statistical offices.

Like Verrijn Stuart, Van Zanten was not opposed to other people undertaking demographic forecasting. In fact, as the person mainly responsible for public statistics, he had an open mind for emerging societal demands. It was he who recognised the quality of Wiebols' forecast and invited him to expound the statistics needed for a municipal population forecast along the methodological lines of his national forecast. He was the first to include a concise description of demographic forecasting methodology and an overview of literature in a manual of statistics for university students (1938). In doing so, this statistician, opposed as he was to the idea of population forecasting in statistics, paradoxically contributed to the anchoring of the new paradigm in the study of future population.

Van Zanten made a clear distinction between forecasts for scientific purposes and for policy making. In his view, Wiebols' forecast was most suitable for scientific purposes, because of its demographic analytical foundation. In purely scientific issues, the speculative character of calculations of future population only played a minor part. On the other hand he was fervently opposed to the application of models like those of Wiebols for policy making purposes. Policy makers could easily be misled by the sophistication of such forecasts and be tempted to forget that the results on which they were going to build their policies were mere speculations. If forecasts were to be made at all, they should be as simple as possible. In that case they could easily be modified when future changes of the tendencies of the vital rates demanded it.

Verrijn Stuart and Van Zanten adhered to the old paradigm of the study of future population in statistics which allowed only for certainty, not for speculation or probability. Methorst, in contrast, even though he did not belong to the mathematical school of statistics, was definitely an advocate of demographic forecasting. In comparison with Verrijn Stuart and Van Zanten, Methorst was the most internationally oriented. He was an outstanding organiser and strategist, able to manoeuvre his international statistical organisation through the vicissitudes of two world wars. He organised the demographic community of the Netherlands as well. The demographic community of the Netherlands owes him a considerable debt.

Methorst was a lawyer by training and lacked a sound background in mathematics. He appears to have been eager to safeguard his status as a man of science amidst the most outstanding public and mathematical statisticians of the world. If he had a weak spot, it was this. He was always conscious of his position among his peers; he avoided embarking on adventures that might put his status among international colleagues in jeopardy. From that point of view it was prudent of him not to intervene in debate with eminent mathematicians and statisticians. He preferred to refer in his publications to his peers, in particular his international counterparts, rather than to the true innovators of forecasting methodology. When demographic forecasting was close to attaining acceptance as the new standard of population forecasting methodology in national and international settings, Methorst's national renown as a statistician enabled him to make many compatriots aware of the significance of age structure for future population growth. He did so in a publication (in 1937) which omitted any reference to the Dutch pioneers of national and municipal demographic forecasting.

In the international context, until the end of the 1920s at least, Methorst was faced with a situation in which the issue of methods in the study of the future of population was unsettled. He had to deal with the uncomfortable problem of remaining consistent in his views with respect to demographic forecasting and yet not taking sides as long as the debate on method was not yet crystallised.

As soon as it re-emerged in the 1920s in the Netherlands, England and elsewhere, demographic forecasting entered a difficult decade. First, a renaissance of the belief in a mathematical law of population growth, namely the law of logistic growth of total population, had to be faced. This law replaced the Malthusian belief in a law of geometric population growth. Because of its origin in biology and the mathematical equation in which it found its expression, logistic growth methodology had all the appearances of a law founded in 'True Science', further demonstrated by its perfect fit of observed curves. It was supported not only by biologists, but also by such eminent mathematical statisticians as Yule. At the annual meeting of the Royal Statistical Society of 1924 and at the first World Population Conference of 1927, Bowley's –speculative– demographic forecasting method got no footing and its advocates almost succumbed in the debate on methods of forecasting future population. It was only at the ISI Tokyo conference in 1931 that demographic forecasting emerged triumphant.

It is understandable that Methorst, with no training in mathematics, did not want to risk his status as an academic, scientist or statistician by intervening in a debate in which mathematical statisticians of international repute crossed swords. Methorst put forward his preference for making forecasts on the basis of extrapolations of the crude rates of the vital components of demographic growth openly, but not in the debating arena. On international scientific occasions he limited himself to statements explicating his view of the dynamics of the course of future population away from the debating halls and without referring to participants in the debate, or contributors to the emergence of demographic forecasting and its innovation. He advocated age-sex specific forecasting, but only after cohort component forecasting was well on its way to becoming the new standard forecasting method.

The national population forecasters

The attitude of the 'internationals' towards demographic forecasting and the positions they took in debates were determined by their restricted view of statistics or by a reluctance to take sides in situations where the outcome of the debate on methodology had not yet crystallised into a generally accepted new 'paradigm'.

But what of the innovators of forecasting methodology themselves? Why did these men fail to take care of the international propagation of information about their innovative endeavours?

The answer is that to do such a thing was beyond their scope. In the Netherlands the debate about the method of calculating future population was never a debate between adherents of the belief in geometrical or logistic laws of population and of speculative demographic forecasting. The great method debate in the inter-war period was between competing demographic methods. The Dutch national population forecasting scene was different from the international scene. Dutch national population forecasts were made by private persons with an interest in the population problem. No forecasts were made by such government bodies as the NCBS, nor by municipal statistical offices. Neither was the innovation of national forecasts directed by the demands of political decision making.

The most innovative forecasters were almost without exception relative outsiders in the field of the scientific study of the population. Wiebols failed to get the position he strove for. 't Hooft did not aspire to such, consciously enjoying being an outsider. For these men, publishing in international scientific journals was out of the question; their main interest was to contribute to the national population debate. Lacking professional or academic status, they could not be delegates at international conferences where the future of population was discussed. Neither could they be asked to represent the Netherlands at international (governmental) occasions where the future of national populations and forecasting methods were discussed.

The situation in the Netherlands differed from that of other European countries. The international contributors to demographic forecasting in the inter-war period whose names do figure in contemporary overviews and international intellectual histories were the chiefs of (national) statistical offices, like Gini, Jensen and Ptoukha, or men with status in the academic or scientific world, like Fahlbeck, Westergaard, Bowley, Yule, Pearl, and Sauvy. The international contributors were on an equal footing with men of the stature of Verrijn Stuart, Methorst and Van Zanten, not Wiebols and 't Hooft.

The innovators of urban and regional forecasting

There was another group of actors, whose position has to be discussed here. Why did the men, who applied national forecasting methodology within the context of municipal and regional planning, fail to penetrate international histories of 20th century population forecasting? It would seem surprising that the contributions to demographic forecasting of Dutch town planners of the 1930s failed to get through to the international histories of population forecasting, for the Dutch planning community took an active part in the international field of (town) planning. These men had their own international contacts, why did they omit to disseminate information about their own and their compatriots' achievements?

The failure is understandable, however, when it is realised that the professional worlds of planners (national and international) and those of statisticians engaged in the quantitative study of the future of population were separate worlds in the period prior to World War II.

Urban and regional planners of the inter-war period were not engaged in the study of the future size of population from mere scientific interest. For them, the demands of town planning and housing were the point of departure. These men had to be open to innovation. Urban planners had to find answers to questions concerning the consequences of future national population increase for their respective municipalities. Such questions concerned the demographic and socio-economic foundations of urban extension plans, or future housing demand. Town planners who focused on these socio-economic foundations belonged to the relatively young profession which was to develop into modern environmental planning.

Without exception, town planner-forecasters like Angenot, Delfgaauw and Van Lohuizen, were intent on judging and testing the utility of innovations of national forecasting methodology (by Wiebols and 't Hooft) for practical urban forecasting and planning purposes. They were eager to share information with colleagues from other countries, but the exchange of information remained restricted to the international town planning forums. Information about innovative developments in demographic forecasting in planning and housing did not reach the international forums of statisticians and demographers, where the study of the future of populations in relation to the population problem was the issue at stake. These two worlds did not mix.

There was yet another impediment to ready propagation of information about innovations at local and regional levels. Most (town) planners were primarily interested in the design of the plans, in the way the plans were based on preliminary town planning research (social surveys), and only secondarily in the specifics of the demographic forecasts underlying these foundations. Their lack of interest in the specifics of demographic forecasting is mirrored in contemporary international planning literature.

9.5 | A Change of Paradigm or a Kuhnian Scientific Revolution?

It was the speculative aspect of the new methodology of demographic forecasting, not the analytic demographic conceptualisation of (future) population dynamics, that induced the early international innovators of population forecasting to mitigate the significance of their demographic futures. The indications for this conclusion, presented in section 9.3 are many and convincing.

Fahlbeck underlined the fact that his extrapolation of the graphs of the birth and death rates was merely a guess. Westergaard called his demographic future a 'horoscope'. Cannan had no desire to stake his reputation as a prophet on the growth of population.

These remarks were made in a climate wherein representatives of mainstream population statistics had great difficulty to renounce the belief in the existence of laws of population or mortality.

In the 1920s logistic growth methodology gave a new stimulus to the waning law-of-population-growth belief. This was the more so because it was introduced by scientists with a Science background, the biologist Pearl and the mathematician and statistician Yule (who adhered to the mathematical school of statistics). In the debate on the merits of logistic growth versus demographic component methodology the respective advocates Yule and Bowley defended themselves against accusations of having tried to make prophesies or predictions. It did not refrain Yule from concluding that the logistic law of population growth is quite capable of representing the growth of a population over what is humanly speaking a fairly long period of time. Also, that the growth of population is a biologically self regulating process, that surpasses the limit of subsistence of an individual nation.

In order to understand the impact of the revitalised belief in a law of population, one should not consider the positions of the mathematicians- statisticians in the first place. Because these men were able to understand the mathematical foundations of the logistic method they could take a relativist position. How did the re-awakened belief effect people with a less thorough mathematical training? The apex of the (revitalised) belief in a law of population was reached in the 1920s, with Thomas Adams as the most outspoken protagonist. In his opinion the law would limit the personal element, the influence of the judgement of the forecaster, on the calculated future.

Outside the Netherlands, at national and international conferences of statisticians and students of the population problem the odds were against demographic forecasting methodology in the 1920s. However, at the Tokyo session of ISI at beginning of the 1930s the tide had turned almost completely. As long as demographic forecasting had not been crystallised into the new mainstream approach, Methorst, the most important representative at these international occasions, refrained from engaging openly in the international debate on the calculation of the future of population between mathematical statisticians and (other) representatives of True Science. It is likely that he abstained from participation because he was afraid of endangering his stature as a prominent statistician.

All the indications mentioned point in one direction: The conclusion that a paradigm shift in a Kuhnian sense had taken place in the study of future population, resulting in a relatively sudden, if not revolutionary change of standard forecasting methodology and underlying belief in the working of Nature in population.

However, there are explanations that do not necessitate the conclusion of a Kuhnian paradigm shift, and there are other indications that contradict the conclusion. For instance, the period of time it took demographic forecasting to re-emerge after a lapse of thirty years since Cannan's forecast, can also be explained from the continuation of the decline of the birth rate and the downward trend of fertility, making extrapolation less hazardous and speculative. The development of public statistics, resulting in an increase of detailed population statistics, and the lengthening of time-series of observed population data made it possible to develop an age (or cohort) specific component approach (Cannan, 1895; Westergaard, 1908; Oly and Bowley, 1924; Wiebols 1925; 't Hooft, 1926) or a less elaborated component approach based on the extrapolation of times series of birth and death rates (e.g. Fahlbeck, 1905; Methorst, 1922).

In the scientific community of students of (the future of) population there is no clear cut professional dividing line between adherents of the old and the new paradigm of the study of future population. Advocates and opponents of the new paradigm can be found among mathematical statisticians (e.g. Bowley,

Holwerda and Westergaard versus Yule), public statisticians (e.g. Gini, Jensen and Methorst versus Verrijn Stuart and Van Zanten) and planners (De Casseres versus Adams). In some instances the contributors to the modernisation of population forecasting came from outside the field of statistics (Cannan and 't Hooft). Others, however, were statisticians themselves (Fahlbeck, Westergaard, Methorst, Bowley, and Wiebols) or were actuaries (Holwerda and Oly).

The Netherlands is a separate case: There was never a debate between advocates of the old and the new paradigm in the 1920s and 1930s. From the moment of its emergence in the early 1920s demographic forecasting had come to stay. However, the most innovative contributions did not come from members of the establishment of public statistics (with the exception of Methorst) but from relative outsiders (Holwerda, Oly, Wiebols, 't Hooft, Van Lohuizen, Delfgaauw, and Angenot). During the 1920s the most prominent members of the establishment (Verrijn Stuart, Van Zanten, and Methorst) did little to further the elaboration of demographic forecasting methodology or the international propagation of knowledge of the state-of-the-art of demographic forecasting in the Netherlands. In contrary. As late as the second half of the 1930s Methorst (by stressing the importance of the age composition in forecasting) and Van Zanten (by devoting a section of a new edition of his textbook on statistics to demographic forecasting) completed the process of demographic forecasting becoming the new standard method and the new paradigm of the study of thc future of population.

If the focus is strictly on methodology the assumption of a revolutionary paradigm shift in a Kuhnian sense is further weakend. The assumption presupposes increasing contradictions between the new and the old method. It is a small step from the total rate of population to a subdivision into a natural growth rate (subdivided into birth rate minus death rate) and a migration surplus rate. Morever, total population calculated with geometrical or logistic grwoth methodology can easily be distributed over age and sex groups, using either a constant or dynamised distribution function. In 1932 Lotka demonstrated how logistic growth methodology and stable population theory could mathematically be linked. When this way of calculating future population by age and sex starts to be unsatisfactory, it is a small step to start thinking in terms of cohorts (and cohort-component methodology). Kersseboom and Cannan had indicated the way. Demographic methodology gave also a new form to the rate of total population growth, which played a central part in the old paradigm. Cohort-component methodology gave rise to stable population theory, which allowed Lotka to calculate the intrinsic growth rate ("Lotka's r"). It is a perfect demonstration of the integration of the old and the new methodology.

Weighing all the arguments, one can only arrive at the conclusion that the transition to modernity of population forecasting was the result of a change of paradigm, although not always a gradual change, rather than a scientific revolution in the community of students of future population. Within the very small Dutch community of official statisticians, however, the change of paradigm must almost have taken the character of a revolutionary transition.

9.6 | The New Position of the Forecaster after the Change of Paradigm

The change of paradigm itself had far reaching consequences for the position of the forecaster. The ultimate consequences, perceived by only a few in the inter-war years, are still resounding.

The consequences have been demonstrated by considering two contemporary reflections on population forecasting dating from the second part of the 1920s. The reflections stem from the regional planners Thomas Adams and Joël de Casseres. Thomas Adams was one of the most outspoken advocates of the old paradigm. He had based his survey of New York and Environs on a logistic population growth approach. He was fascinated by the logistic law of population growth, because the future course of the curve of total population is determined by the past and not by the judgement of the forecaster. He hoped and believed that a law of population growth would reduce the play of the forecaster's personal judgement. Adams saw the position of the forecaster as that of a neutral, objective outsider: External to the population system and not a part of it. In his view the forecaster's only task is to find the right values for the parameters of the mathematical logistic growth equation. An opposite view, coinciding with the new paradigm, was taken by Adams' Dutch counterpart, Joël de Casseres. De Casseres stressed the fact that mathematical conceptualisations such as logistic growth were not different from other conceptualisations helping mankind to come to terms with (future) reality. The image of the future is blurred; it can only be captured with conceptualisations of reality that are the product of the human mind. As Kuczynski put it at the end of the 1930s, demographic forecasts are at best reasoned estimations.

The theoretical philosophical foundations of the new situation can be found in George Herbert Mead's philosophy of the present, dating also from the 1930s. Meads' premise was that the locus of reality is the present. The present can only be perceived by what are called 'emerging' events. The fundamental characteristic of an emerging event is that it has an aspect of novelty, of unexpectedness. Emerging events break into time. Without them, the present can not be perceived. Once an event has emerged, it is up to the human mind to establish a new continuity of time, a past, into which the event 'fits'. The past is a mental reconstruction; it is a re-established continuity. Once continuity is re-established it is fit for extrapolation into the future. Such symbolically reconstructed pasts and re-established continuities stand in opposition to 'social structural pasts' (in Meadian philosophy). Social structural pasts are the pasts Thomas Adams and many like minded forecasters feel most comfortable with. Social structural pasts structure and condition the events that emerge in the present and enable to start from the premise of causality in stead of the interference of the forecasters mind.

The fundamental implication of these and other dimensions of the past in the philosophy of Mead is that the position of the forecaster has to come to the fore. The forecaster is not just the person responsible for Kuczynski's 'reasoned guess'. As the ultimate consequence of the paradigm change in population forecasting, the former position of the forecaster as the neutral and objective outsider has turned into something else. The outsider in Adams's forecasting system has become an inherent part of the system to be forecast. The consequence of the changed position is that current long-term societal expectations, hopes and fears with respect to the future of population, whether inspired by Malthus, social Darwinism or others, are reflected in the forecasters' assumption making.

In the course of time, other unwelcome insights forced themselves onto the forecasting profession. For a while, particularly in the 1930s, the forecaster remained primarily the provider of information of future reality, attempting to be as objective as possible. This has been the preferred view of the forecasting profession itself, and it is also what society wants it to be. The difficulty Dutch forecasters had in departing from the former pre-paradigm change position in society is reflected in the way urban and regional forecasters attempted to cope with the problem posed by migration. The difficulty and the speculative character of forecasting future migration made these men increasingly dissatisfied with demographic forecasting.

In the Netherlands it was only after the Second World War that the openness of the future induced policy makers to attempt to direct migration flows, mainly through regional economic policies and local and regional housing allocations. In the last decades of the 20th century the consequences of the new paradigm for the position of the forecaster have become gradually more pronounced.

In a way, the situation of the 1990s resembles the uncertainty Cannan faced a century ago. Then traditional geometrical forecasting of total population, based on the constancy of the crude growth rate, was being subjected to severe criticism because of a trend bend in natality rates. Cannan decided to start from a new concept of reality by constructing an explanatory demographic analytical model of population dynamics instead, where the cohort structure of the population played an important part. Now, at the turn of a new century, mere extrapolation of observed trends of occurrence/exposure rates of the vital components of population growth in the present highly sophisticated forecasting models has become as hazardous as in Cannan's days, for instance because of the extremely low levels fertility rates have reached. Nowadays, no forecaster of national, regional or urban population could allow himself to exclude migration from the calculations. As a consequence of the continuing process of individualisation and lifestyle developments a segmentation of buyers markets (demand markets) takes place. This asks for a new conceptualisation and modelling of that part of socio-economic reality in which demographic processes take place. Life event history analysis, the fruit of the new paradigm in demography, as Willekens and Courgeau and Lelièvre see it (Section 1.4) may well provide an answer in due time.

If a new conceptualisation were to be made of socio-economic and demographic reality, it was explicitly to include the forecaster as part of the forecasting process and as part of the system that is forecast, for instance in terms of target forecasting. In target forecasts the political choices are made explicit.

Such a new concept and model of population dynamics could help save population forecasting as a serious, credible activity. Nowadays, population forecasting as an activity directed at creating points of orientation for public and private decision making is at risk. Increased uncertainty with respect to the future is revealed by the frequency with which forecasts are being made and the popularity of monitoring, (micro and macro) simulation and scenario methodology. Population forecasting is at risk because the position of forecasters continues to change at a fast rate. Population forecasting is at risk in all instances where a protective screen between the desires and demands of authorities commissioning forecasts and the forecaster is lacking. Forecasters are more than ever before running the risks of intimate involvement and vested interests in the futures that are forecast. Government bodies at urban, regional and national levels are increasingly contracting out the exploration of demographic futures and resulting consequences for housing, schooling, health, financing of retirement and so forth to private agencies and groups of university researchers. These groups are engaged in vehement market competition, competition for economic survival, or for influence on the centres of power and decision making. In negotiations between government bodies (for example, between municipalities and the state) and between local authorities commissioning forecasts and forecaster, balances of power play an important part. The quality of the information provided by negotiators, and by implication by the forecaster, is judged on its utility in the negotiation process, speed of production, effectiveness, credibility and acceptability for the commissioner. The operation of the forecasting market creates new demands and sets new rules.

Because of their dependency of commissions or need to operate near to the centres of power, these competitors in the forecasting market depend increasingly on the financial providers of forecasting commissions and the satisfaction of those who make the commission. The dependency could threaten the quality of demographic futures research and the credibility of population forecasts.

A hundred years have passed since demographic forecasting first emerged. It was the manifestation of a change of paradigm in the study of future population. Demographic analysis, population theory and speculation about the future course of population have replaced a scientific 'law' and certainty. Understanding of population dynamics has increased enormously, as has the sophistication of its modelling. It was Cannan's genius to create new trust in population forecasting by the introduction of the demographic momentum into the study of future population, in terms of its age/cohort structure. It was Westergaard who demonstrated how demographic insight into population dynamics could be used for a concise, imaginative description of the demographic transition European populations would encounter in the 20th century and for an evocation of its consequences. It was Wiebols, Angenot, Delfgaauw and Van Lohuizen who indicated how the new insights could be adequately and accurately modelled for forecasting purposes and planning practice. It was De Casseres who demonstrated most clearly the consequences of the new paradigm for the position of the forecaster. The dilemma that it posed to the community of statisticians was most clearly analysed by Van Zanten: Even though statisticians could be assumed to be the best equipped profession for making population forecasts, they should nevertheless abstain from forecasting, to safeguard the reliability of public statistics, although they should provide good data for the demographic forecasting purposes of others.

At present, the population forecasting profession is faced with a similar challenge, namely how to retain society's trust in population forecasts. The challenge can be postponed by focusing on the construction of a life event history based forecasting model, or can be circumvented by focusing on futures simulations, on scenario methodology, or in increasing frequency of forecasts and an emphasis on monitoring.

The challenge might as well be accepted. In that case, the condition has to be fulfilled that a protective screen be erected between the forecaster and all conscious and purposeful attempts to influence his results, those who commissioned the task of making a forecast in the first place. With that condition fulfilled, the forecaster can proceed to make target projections and be explicit about the political and target related choices constituting the basis of the forecast.

References

Adam, B. (1990), *Time and social theory*. Cambridge: Polity Press.

Adams, Th., H.M. Lewis, and Th.T. McCrosky (1929), Population, landvalues and government. Studies of the growth, distribution of population and landvalues; and of problems of government. Regional survey, Vol. II. *Regional plan of New York and its environs*. New York

Alberts, G. (1989), Introduction. The relations between the study of mathematics in the societal context. In: G. Alberts, H.J.M. Bos, and J. Nuis, *Om de wiskunde. Stimulansen voor toepassingsgerichte wiskunde rond 1946*. Amsterdam: Centre for Mathematics and Computer Science, pp. 1-6.

Amsterdam. *Journal of the town planning institute*, Vol. XXI (1935) 10, pp. 271-273.

Angenot, L.H.J. (1934), *De toekomstige loop der bevolking in Nederland en in het havengebied van Rotterdam*. Met een inleiding van Ir. W.G. Witteveen. Rotterdam: Vereeniging voor Stadsverbetering "Nieuw-Rotterdam".

Angenot, L.H.J. (1935), Over de toekomstige omvang van de woningvoorraad in Nederland. *Tijdschrift voor Volkshuisvesting en Stedebouw* , Vol. 16, nr. 11, pp. 196-203.

Angenot, L.H.J. (1973), Schets van de eerste kwart-eeuw stedebouwkundig onderzoek in Nederland. *Tijdschrift voor Volkshuisvesting en Stedebouw*, Vol. 54, pp. 88-95.

Ankersmit, F.R. (1990), Over geschiedenis en tijd. In: F.R. Ankersmit, *De navel van de geschiedenis. Over interpretatie, representatie em historische realiteit*. Groningen: Historische Uitgeverij Groningen, pp. 109-124.

Ascher, W. (1978), *Forecasting. An appraisal for policy-makers and planners*. Baltimore.

Bakker Schut, P. (1915), Bevolkingsbeweging en woningvoorziening in onze groote steden. *De Ingenieur*, Vol. 30, nr. 34, pp. 693-691.

Bakker Schut, P. (1921), De resultaten der woningtelling I. *Gemeentebelangen*, Vol. 16, nr. 21, pp. 324-326.

Bakker Schut, P. (1933), *De bevolkingsbeweging in Nederland en in het bijzonder te 's-Gravenhage*. Alphen a/d Rijn: N. Samson N.V.

Bakker Schut, P. and F. Bakker Schut (1944), *Planologie. Van uitbreidingsplan over streekplan naar nationaal plan*. Gorinchem: J. Noorduijn en Zoon N.V.

Baumeister, R. (1876), *Stadt-Erweiterungen in technischer, baupolizeilicher und wirtschaftlicher Beziehung*. Berlin: Verlag von Ernst und Korn.

Baumgartner, Th. (1987), *Population forecasting, policy-making and socio-political context. Lessons to be learnt from 20 years of energy forecasting*. Paper prepared for presentation at the Session on Population Forecasting and Planning, European Population Conference 1987, Jyväskylä, Finland, June 11-16, 34 pp.

Bebouwing en behoud van natuurruimte. Rapport uitgebracht door een commissie ingesteld door de Commissie voor den Monumentendag en het Nederlandsch Instituut voor Volkshuisvesting en Stedebouw (1942). Alphen a/d Rijn: N. Samson N.V.

Becker, H.A. (ed.) (1992), *Dynamics of cohort and generations research*. Proceedings of a symposium held on 12, 13 and 14 December 1991 at the University of Utrecht, The Netherlands. Amsterdam: Thesis Publishers.

Benser, H. (1940), Vorausberechnungen der Bevölkerungsentwicklung. In: Fr. Burgdörfer (ed.), *Die Statistik in Deutschland nach ihrem heutigen Stand*. Ehrengabe für Friedrich Zahn. Berlin.

Bergmann, W. (1992), The problem of time in sociology: An overview of the literature on the state of theory and research on the 'Sociology of Time', 1900-1982. *Time and Society*, Vol. 1, nr. 1, pp. 81-134.

Bertillon, J. (1903), *Rapport sur les relations entre la mortalité et la natalité, dans les différents pays de l'Europe et notamment en France*. Paris.

Bierens de Haan jr., J. (1924), *Het fatum van de bevolkingsvermeerdering*. Haarlem: Erven F. Bohn.

Bloemers, H.P.J. (1920), De financieële zijde van het woningvraagstuk. *Vragen des Tijds*, Vol. 46, pp. 430-455.

Boelens, P.L.R. (1994), *The state of the art in population forecasting in the former Soviet Union during the 1920s and 1930s*. Assignment PDOD-Course population and household forecasting. Population Research Centre, Groningen University. Unpublished.

Bolte, K.M and D. Kappe (1967), Struktur und Entwicklung der Bevölkerung. In: *Struktur und Wandel der Gesellschaft*, Reihe B der Beiträge zur Sozialkunde nr. 2, Opladen: Leske Verlag.

Bosma, K. (1990), De eerste planoloog. J.M. de Casseres (1902-1990), *Archis*, nr. 10, pp. 5-6.

Bosma, K. (1993), *Ruimte voor een nieuwe tijd. Vormgeving van de Nederlandse regio 1900-1945*. Rotterdam: NAI.

Bouthoul, G. (1935), *La population dans le monde. Les grands événements historiques. Guerre et population. Perspectives d'avenir*. Paris: Payot.

Bowley, A.L. (1924), Births and population in Great-Britain. *The Economic Journal*, Vol. 34, pp. 188-192.

Bowley, A.L. (1935), Obituary Edwin Cannan. *The Economic Journal*, Vol. XLV, 77, pp. 385-398.

Bruyn, W.J. (1976), *Delegatie bij planvoorbereiding. Een kritische beschouwing over het Algemeen Uitbreidingsplan voor Amsterdam, 1935*. PhD Thesis Catholic University of Tilburg.

Burgdörfer, Fr. (1935), *Aufbau und Bewegung der Bevölkerung. Ein Führer durch die Deutsche Bevölkerungsstatistik und Bevölkerungspolitik*. Leipzig.

Burgdörfer, Fr. (1935/36), Zur Frage der bereinigten Lebensbilanz und der Vorausberechnungen über die Bevölkerungsentwicklung. Bemerkungen zu den Vorstehenden Ausführungen van Dr. Paulinus. *Allgemeines Statistisches Archiv* (Jena: Verlag Gustav Fischer), Vol. 25, pp. 155-165.

Cannan, E. (1894), The growth of Manchester and Liverpool, 1801-1891. *The Economic Journal*, Vol. 4, nr. 19, pp. 111-114

Cannan, E. (1895), The probability of a cessation in the growth of population in England and Wales during the next century. *The Economic Journal*, Vol. 5, nr. 20, pp. 505-515.

Cannan, E. (1931), The changed outlook in regard to population, 1831-1931. *The Economic Journal*, Vol. 41, nr. 164, pp. 519-534.

Carr-Saunders, A.M. (1925), *Population*. London.

Carr-Saunders, A.M. (1936), *World population. past growth and present trends*. Oxford: Clarendon Press.

Caselli, G. and E. Lombardo (1990), Graphiques et analyse démographique: quelques éléments d'histoire et d'actualité. *Population* (édition de l'INED), nr. 2, pp. 399-414.

Casti, J. L. (1989), Paradigms lost. Images of man in the mirror of science. New York: William Morrow and Comp.

CBS (1937), Het probleem van de toekomstige bevolkingsgrootte van Nederland. *Mededeeling*, nr. 3, Maart 1937.

CBS (1951), Berekeningen omtrent de toekomstige loop der Nederlandse bevolking. Utrecht: Uitgevers Maatschappij W. de Haan N.V.

CBS (1975), *Generatie-sterftetafels voor Nederland afgeleid uit waarnemingen over de periode 1871-1973*. 's-Gravenhage: Staatsuitgeverij.

Chesnais, J.C. (1996), *The demographic transition. Stages, patterns, and economic implications. A longitudinal study of sixty-seven countries covering the period 1720-1984*. Clarendon Press.

Cliquet, R.L. (1991), De tweede demografische transitie: feit of fictie? *Bevolking en Gezin*, 2, pp. 1-42.

Courgeau, D. and É. Lelièvre (1996), Changement de paradigme en démographie. *Population*, 3, pp. 645-654.

Cramér, H. (1935), Über die Vorausberechnung der Bevölkerungsentwicklung in Schweden. *Skandinavisk Aktuarietidskrift*, Vol. 18, pp. 35-54.

Cruijsen, H. and N.W. Keilman (1984), *Prognose van de bevolking van Nederland na 1980*. Deel 2. Modelbouw en hypothesevorming, NCBS-publications, 's-Gravenhage: Staatsuitgeverij.

David, H.P., J. Fleischhacker and Ch. Höhn (1988), Abortion and eugenics in Nazi Germany. *Population and Development Review* , Vol. 14, nr. 1, pp. 81-112.

De Casseres, J.M. (1926), *Stedebouw*. Amsterdam: S.L. van Looy.

De Gans, H. A. (1989), *De gemeentelijke bevolkingsprognose*. Muiderberg: Coutinho.

De Gans, H.A. (1993), Bevolkingsprognoses voor Nederland in het interbellum. *Bevolking en Gezin*, nr. 1, pp. 65-92.

De Gans, H. A. (1994a), *De kijk van een statisticus ten tijde van het Interbellum op de bevolkingsprognostiek: Mr Dr J. H. van Zanten (1874-1944)*. Amsterdam: PDI/UvA and O+S.

De Gans, H. A. (1994b), Breaking into time? Problems of time in demographic forecasting. In: Éric Vilquin (ed.) (1994), *Le temps et la démographie*. Chaire Quételet 1993. Louvain-la-Neuve: Academia/L'Harmattan, pp. 329-358.

De Gans, H.A. (1995a), Population forecasting in The Netherlands between the two world wars. In: H. van den Brekel and F. Deven (eds.), *Population and Family in the Low Countries 1994. Selected Current Issues*. Dordrecht: Kluwer Academic Publishers, pp. 125-154.

De Gans, H.A. (1995b), Prognoses maken is mensenwerk: effecten van de maatschappelijke context op methoden en resultaten. In R.F. van der Erf, H.E. Gordijn and A.C. Liefbroer, Demografische prognoses in de praktijk. *Bevolking en Gezin*. Boekaflevering 1995, pp. 17-42.

De Gans, H.A. (1996), *Looking back to the future of Europe's population. Lessons to be learnt from early endeavours in long-term population forecasting*. In: *Evolution or revolution in European population*. Vol. 3. Milano: FrancoAngeli, 1996, pp. 243-256.

De Gans, H.A. (1997), *Demographic forecasting in the Netherlands 1895-1945. The analysis and implications of a paradigm shift*. PhD thesis University of Amsterdam. Amsterdam.

De Jong, A.M. (1946), *Inleiding tot het bevolkingsvraagstuk*. 's-Gravenhage: Martinus Nijhoff.

De Jonge van Ellemeet, M.J.I. (1932), Voorwoord. In: Nederlandsch Instituut voor Volkshuisvesting en Stedebouw, Serie voor Stedebouw, nr. VII.

Delfgaauw, G.Th.J. (1932), *'De tendenzen tot decentralisatie in the vestiging der nijverheid' en 'De omvang van het forensenwezen in Nederland'*. Nederlandsch Instituut voor Volkshuisvesting en Stedebouw, Serie voor Stedebouw VII, pp. 1-41; pp. 43-64.

Delfgaauw, G. Th. J. (1933A), Boekbespreking Ir.P. Bakker Schut: "De bevolkingsbeweging in Nederland en in het bijzonder te 's Gravenhage". Alphen aan den Rijn 1933. *Tijdschrift voor Volkshuisvesting en Stedebouw*, Vol. 14, pp. 194-195.

Delfgaauw, G. Th. J. (1933B), A study of future population growth in Amsterdam. *Journal of the Town Planning Institute*, Vol. XIX, nr. 4, pp. 79-80.

Delfgaauw, G. Th. J. (1935), Boekbespreking Ir. L.H.J. Angenot, De toekomstige loop der bevolking van Nederland en het havengebied van Rotterdam; Rotterdam 1934. *Tijdschrift voor Volkshuisvesting en Stedebouw*, Vol. 16, pp. 179-180.

Denbigh, K.G. (1981), *Three concepts of time*. Berlin/Heidelberg/New York: Springer Verlag.

Derksen, J.B.D. (1937), Juiste formules in het bevolkingsvraagstuk. *De Economist*, Vol. 86, pp. 281-285.

De Ruijter, P. (1979), Ir J.M. de Casseres, stedebouwer en planoloog. *Rooilijn*, nr. 4, pp. 3-10.

De Ruijter, P. (1987), *Voor volkshuisvesting en stedebouw*. Utrecht: Stichting Matrijs.

De Vos, S. and A. Palloni (1989), Formal models and methods for the analysis of kinship and household organization. *Population Index*, Vol. 55, nr. 2, pp. 174-198.

De Vries, G. (1985), *De ontwikkeling van wetenschap. Een inleiding in de wetenschapsfiloso-fie*. Groningen: Wolters-Noordhoff.

De Vries, F. (1948), Prof. dr. C.A. Verrijn Stuart, 22 December 1865 - 26 October 1948. *De Economist*, Vol. 96, pp. 721-742.

Diels, A.E. (1951; 1953), *Opvattingen van ondertrouwde vrouwen omtrent de grootte van haar toekomstig gezin*, Vol. I (1951); Vol. II (1953). Amsterdam: ISONEVO/Commissie voor het Geboorte-Onderzoek.

Discussion on Dr. Stevenson's paper, *Journal of the Royal Statistical Society*, Vol. 88 (1925), pp. 76-90.

Dougill, W. (1936), Amsterdam general extension plan. *The Town Planning Review*, Vol. XVII, nr. 1, pp. 1-10.

Dumont, A. (1890), *Dépopulation et civilization*. Paris.

Dupâquier, J. and M. (1985), *Histoire de la démographie. La statistique de la population des origines à 1914*. Paris: Perrin.

Duparc, H.J.A. and A.W. Grootendorst (1978), Historical survey. In: E.M.J. Bertin, H.J.M. Bos and A.W. Grootendorst, eds., *Two decades of mathematics in the Netherlands. 1920-1940. A retrospection on the occasion of the bicentennial of the Wiskundig Genootschap*, Amsterdam: Mathematisch Centrum, pp. v-xxiv.

Elias, N. (1982/1985), *Een essay over tijd*. Amsterdam: Meulenhoff Nederland B.V.

Engberts, G.E. (1973), *Woningmarkt- en woningbehoefteonderzoek. De wederzijdse relatie tussen onderzoek en beleid*. Deventer: Kluwer.

Esenwein-Rothe, I. (1982), *Einführung in die Demographie. Bevölkerungsstruktur und Bevölkerungsprzess aus der Sicht der Statistik*. Wiesbaden: Franz Steiner Verlag GMBH.

Faber, J.H. (1908), Stadsuitbreiding in de Vereeniging voor de Staathuishoudkunde en de Statistiek. *Gemeentebelangen*, Vol. 4, nr. 6, pp. 88-90.

Faber, J.H. (1911), Iets over uitbreidingsplannen hier en elders. *Gemeentebelangen*, Vol. 7, nr. 3.

Faber, J.H. (1918), Hoeveel woningen? *Gemeentebelangen* 13, 13, pp. 184-186.

Fahlbeck, P. E. (1903), *Der Adel Schwedens*. Jena.

Fahlbeck, P. (1905), La décadence et la chute des peuples. *Bulletin de l'Institut International de Statistique*, Vol. XV, nr. 2, Annexe 12, pp. 367-389. Rome.

Faludi, A. and A. van der Valk (1994), *Rule and order. Dutch planning doctrine in the twentieth century*. Kluwer Academic Publishers: Dordrecht, et cetera.

Friedmann, J. (1973), *Retracking America. A theory of transactive planning*. New York.

Geddes, P. (1915/1949), *Cities in evolution*. Londen.

Giddens, A. (1984), *The constitution of society. Outline of the theory of structuration*. Cambridge: Polity Press.

Giedion, S. (1940/1966), *Space, time and architecture. The growth of a new tradition*. Cambridge, Massachusetts: Harvard University Press.

Gini, C. (1930), The Italian demographic problem and the fascist policy on population. *The Journal of Political Economy*, Vol. 38, pp. 682-697.

Gini, C. (1931A), Calcolo di previsione della popolazione italiana dal 1921 al 1961. *Bulletin de l'Institut International de Statistique*, Tome XXV, 3ème livraison. 's-Gravenhage, pp. 35-40.

Gini, C. (1931B), La XIXe session de l'Institut International de Statistique. *Metron*, Vol. 9, nr. 4, pp. 107-155.

Girard, A. (1958/1968), Sociale demografie. Vraagstukken van sociale morfologie. In: G. Gurvitch, *Handbook van de sociologie* 2, Aula 362. Utrecht/Antwerpen: Het Spectrum, pp. 43-71.

Glass, D.V. (1940/1967), *Population. policies and movements in Europe*. London: Frank Cass and Comp. Ltd.

Godefroy, J. (1960), Sociale prognose. *Sociale Wetenschappen*, Vol. 3 (1960), nr. 2, pp. 183-201.

Greenwood, M. (1925), The growth of population in England and Wales. *Metron*, Vol. 1, nr. 1, pp. 113-125.

Grondslagen voor de stedebouwkundige ontwikkeling van Amsterdam. Studie betreffende de toekomstige bevolkingsaanwas (1932). Amsterdam: Stadsdrukkerij.

Hagendijk, R.P. (1996), *Wetenschap, constructivisme en cultuur*. PhD University of Amsterdam.

Hajnal, J.(1955), The prospects of population forecasts. *Journal of the American Statistical Association*, Vol. 50, pp. 309-322.

Hecht, J. (1980), Imagination et prospective: les origines de la prévision démographique. In: *Quellenabhandlung zur Geschichte der Staatsbeschreibung und Statistik*. I.Band. Statistik und Staatsbeschreibung in der Neuzeit, vornehmlich im 16-18. Jahrhundert. Paderborn, München, Wien, Zurich.

Hecht, J. (1990), l'Avenir était leur affaire: de quelques essais de prévision démographique au XVIIIème siècle. *European Journal of Population*, Vol. 6, pp. 285-332.

Heinsohn, G., R. Knieper, and O. Steiger (1979), *Menschenproduktion. Allgemeine Bevölkerungslehre der Neuzeit.*Frankfurt am Main: Suhrkamp Verlag.

Hellinga, H. (1985), De eerste jaren van de afdeling Stadsontwikkeling. In: H. Hellinga and P. de Ruijter, *Algemeen uitbreidingsplan 50 jaar*. Amsterdam: Amsterdamse Raad voor de Stedebouw.

Hellinga, H. and P. de Ruijter (1985), *Algemeen uitbreidingsplan Amsterdam 50 jaar*. Amsterdam: Amsterdamse Raad voor de Stedebouw.

Henssen, E.W.A. (1994), *Een welmenend cynicus. Opkomst, val en eerherstel van Mr L.A. Ries*. Amsterdam: Lubberhuizen.

Hiller, E.T. (1930), A culture theory of population trends. *The Journal of Political Economy*, Vol. 38, pp. 523-550.

Hodgson, D. (1983), Demography as social science and policy science. *Population and Development Review*, Vol. 9, nr. 1, pp. 1-34.

Hodgson, D. (1991), The ideological origins of the population association of America. *Population and Development Review*, Vol. 17, nr. 1, pp. .

Höhn, Ch. (1986), Amtliche Bevölkerungsvorausschätzungen seit 1925 – eine kurze Geschichte der Politikberatung un des demographischen Klimas. In: K. Hanau a. o.(eds.), *Wirtschafts- und Sozialstatistik. Empirische Grundlage politischer Unterscheidungen*. Heinz Grohmann zum 65. Geburtstag. Göttingen, pp. 209-231.

Holwerda, A.O. (1917), *De wetenschappelijke richting in de statistiek*. Rotterdam.

Holwerda, A.O. (1926), De methode van berekening van het toekomstige bevolkingsaccres. *De Economist*, Vol. 75, pp. 854-866.

Holwerda, A.O. (1927), Hoe het toekomstige bevolkingscijfer van Nederland niet berekend moet worden. *De Economist*, Vol. 76, pp. 148-152.

Holwerda, A.O. (1935), De bevolkingstoeneming van Nederland en van Amsterdam, 's-Gravenhage en Rotterdam in het bijzonder. *Tijdschrift voor Economische Geografie* Vol. 26, nr. 2, pp. 25-30.

Honey, F.J.C. (1937), The estimated population of Great Britain, 1941-1971. *Journal of the Institute of Actuaries*, Vol. 68, pp. 323-347. Abstract of the Discussion, ibidem, pp. 348-368.

Houwaart, E.S. (1991), *De hygiënisten. Artsen, staat en volksgezondheid in Nederland, 1840-1890*. Groningen: Historische Uitgeverij Groningen.

Huygens, Chr. (1920), *Oevres complètes*. La Haye: Martinus Nijhoff, t. 6 (Correspondance), pp. 515-538.

Huijs, H.W.J.M. (1974), *De bevolkingsontwikkeling in Nederland 1929-1940. Een geannoteerde bibliografie van de contemporaine demografische literatuur*. Intern rapport, nr. 2. Voorburg: NIDI.

International Statistical Institute (1985), *The International Statistical Institute (1885-1985)*. Voorburg.

Ipsen, C. (1996), *Dictating demography. The problem of population in fascist Italy*. Cambridge University Press.

Jensen, A.L.O. (1931), Horoscope of the population of Denmark. *Bulletin de l'Institut International de Statistique*, Tome XXV, 3ème Livraison, pp. 41-49.

Joas, H. (1985), *G.H. Mead. A contemporary re-examination of his thought*. Cambridge: Polity Press.

Keilman, N.W. (1990), *Uncertainty in national population forecasting: issues, backgrounds, analyses, recommendations*. Publications of the Netherlands Interdisciplinary Demographic Institute (NIDI) and Family Study Centre (CBGS), Vol 20. Amsterdam: Swets and Zeitlinger.

Kendall, M.G. (1956/1970), The beginnings of a probability calculus. In: E.S. Pearson and M.G. Kendall (eds.; 1970), *Studies in the history of statistics and probability*. London: Griffin. Reprint from *Biometrika* , Vol. 43 (1956), pp. 1-14.

Kendall, M.G. (1960/1970), Where shall the history of statistics begin? In: E.S. Pearson and M.G. Kendall (eds.; 1970), *Studies in the history of statistics and probability*. London: Griffin. Reprint from *Biometrika*, Vol. 47 (1960), pp. 447-449.

Kenessey, Z. (1993), *The establishment of the ISI permanent office in the Netherlands in 1913*. Note prepared on the occasion of the commemoration of the 80th Anniversary, Voorburg, April 27, 1993. Voorburg: ISI.

Kersseboom, W. (1738-1742), *Proeven van politique rekenkunde vervat in drie verhandelingen over de meenigte des volks in de provintie van Hollandt en Westvrieslandt; de probable leeftijd der weduwen, de duurzaamheid der huwelijken, de relatie van de meenigte des volks tegen het getal der geborene, en dat der gehuwde paaren & alsmede over de waardije van lyfrente in proportie van losrente: waarbij gevoegt zijn eenige opmerkingen op de gissingen over den staat van het menschelijk geslacht & begrepen in zeekere inleiding tot de algemeene geographie: eerste verzameling*. 's-Gravenhage: Jan van den Bergh.

Keyfitz, N. (1972), On future population. *Journal of the American Statistical Association*, Vol. 67, nr. 338, pp. 347-363.

Keyfitz. N. (1981), The limits of population forecasting. *Population and Development Review*, Vol. 7, nr. 4, pp. 579-593.

Keyfitz, N. (1990), Demographic forecasting; state of the art and research needs; a comment. In: C.A. Hazeu and G.A.B. Frinking, *Emerging issues in demographic research*. Amsterdam: Elsevier, pp. 67-71.

Kirk, D. (1996), Demographic transition theory. *Population Studies*, Vol. 50, pp. 361-388.

Klaassen, L.H., D.H. van Dongen Torman, and L.M. Koyck (1949), *Hoofdlijnen van de sociaal-economische ontwikkeling der gemeente Amersfoort van 1900-1970*. Leiden: Stenfert Kroese's Uitgevers Maatschappij.

Kloos, W.B. (1939), *Het nationaal plan. Proeve eener beschrijving der planologische ontwikkelingsmogelijkheden voor Nederland*. Alphen aan den Rijn: N. Samson N.V.

KNAW (Koninklijke Nederlandse Akademie van Wetenschappen; 1966), *De Mens in dichte pakking*. Symposium gehouden op maandag 20 en dinsdag 21 juni 1966 te Amsterdam in het Koninklijk Instituut voor de Tropen. Amsterdam: N.V. Noord-Hollandsche Uitgevers Maatschappij.

Kohlbrugge, J.H.F. (1922), De ethische zijde van het neo-Malthusianisme. *Economisch-Statistische Berichten*, Vol. 7, pp. 941.

Kohlbrugge, J.H.F. (1929), *Practische sociologie. Deel VI: Sociale nooden II. Sexueele en geestelijke problemen*. Groningen-Amsterdam: J.B.Wolters Uitgevers-maatschappij N.V.

Krausz, E. (1986), *Sociological research. A philosophy of science perspective*. Assen/-Wolfeboro: Van Gorcum.

Kronjee, G. (1992), De dubbele werkelijkheid van het gezin. *Gezin. Tijdschrift voor primaire leefvormen*, Vol. 4, nr. 3-4, pp. 157-170.

Kuczynski, J. (1972), *Studien zur Wissenschaft von den Gesellschaftswissenschaften. Erinnerungen, Erfahrungen, Überlegungen*. Berlin: VEB Deutscher Verlag der Wissenschaften.

Kuczynski, R.R. (1928), *The balance of births and deaths*, Vol. I, *Western and Northern Europe*. New York: Macmillan.

Kuhn, Th. S. (1972/1970), *De structuur van wetenschappelijke kennis*. Meppel: J.A. Boom en Zn.

Kuijsten, A.C. (1994), Buitenechtelijke geboorten in Nederland: een demo-geografische analyse. *Acta Geographica Lovaniensia*, Vol. 34, 1994, pp. 23-41.

Kuijsten, A.C. (1996), *Beatrijs en de kip*. Amsterdam:AME.

Lachiver, M. (1987), Les projections avant l'époque contemporaine. In: *Les projections démographiques*. Actes du Colloque, Tome 1. Traveaux et Documents, Cahier no.116, Paris: P.U.F., pp. 29-39.

Landry, A. (1949), *Traité de démographie*. Paris: Payot.

Le Bras, H. (1987), Nature et limites des prévisions de population. In: *Les projections démographiques*. Actes du Colloque, Tome 1. Traveaux et Documents, Cahier no. 116, Paris: P.U.F., pp 13-26.

Leidraad bij het samenstellen van uitbreidingsplannen (1927). Vaste Commissie voor de Uitbreidingsplannen in Noord-Holland. Haarlem.

Lesthaeghe, R. and D.J. van de Kaa (1986), Twee demografische transities? In: D.J. van de Kaa and R. Lesthaeghe (eds), Bevolking: groei en krimp. *Mens en Maatschappij*, pp. 9-24.

Lewandowski, H. and P.J. van Dranen (1933), *Beschavings- en zedengeschiedenis van Nederland*. Amsterdam: N.V. Uitgevers Maatschappij ENUM.

Lewes, F.M.M. (1984), A note on the origin of the net reproduction ratio. *Population Studies*, Vol. 38, pp. 321-324.

Lorenz, C. (1987), *De constructie van het verleden. Een inleiding in de theorie van de geschiedenis*. Meppel/Amsterdam: Boom.

Lotka, A.J. (1932), The structure of a growing population. Paper no. 20 in G.H.L.F. Pitt-Rivers, *Problems of Population,* pp. 261-297.

Lowenthal, D. (1985), *The past is a foreign country*. Cambridge: Cambridge University Press.

Lutz, W., W. Sanderson, and S. Scherbow (1997), Doubling of world population unlikely. *Nature*, Vol. 387, 19 June 1996, pp. 803-805.

Lynch, K. (1972), *What time is this place?* Cambridge (Mass.) and London (England): the MIT Press.

Maines, D.R. (1978), Bodies and selves: Notes on a fundamental dilemma in demography. In: N. Denzin (ed.), *Studies in Symbolic Interaction*, Vol. I. Greenwich, CT: JAI Press, pp. 241-265.

Maines, D.R., M.A. Katovich, and N.M. Sugrue (1983), The sociological import of G.H.Mead's theory of the past. *American Sociological Review*, Vol. 48, nr. 2, pp. 161--173.

Malthus, Th. (1798/1870; 7th ed.), *Essay on the principle of population*.

Masterman, M. (1970/1974), The nature of a paradigm. In: I. Lakatos and A. Musgrave (eds.), *Criticism and the growth of knowledge*. London: Cambridge University Press, pp. 59-89.

Mead, G.H. (1929), The nature of the past. In: John Coss (ed.), *Essays in honor of John Dewey*. New York: Henri Holt, pp. 235-242.

Mead, G.H. (1932/1980), *The philosophy of the present*. Chicago/London: The University of Chicago Press.

Mercure, D. (1983), Typologie des représentations de l'avenir. *Loisir et société/society and leisure*, Vol. 6, nr. 2, pp. 375-404.

Methorst, H.W. (1895), *Werkinrichtingen voor behoeftigen*.

Methorst, H.W. (1914), *De beteekenis van den achteruitgang van het geboortecijfer in Nederland*. Amsterdam: Maatschappij tot Nut van 't Algemeen.

Methorst, H.W. (1919), Bevolkingsvraagstuk. In: *Handelingen van het Nationaal Congres tegen het nieuw-Malthusianisme en ter bespreking van het bevolkings-vraagstuk*. Arnhem, pp. 77-87.

Methorst, H.W. (1922), Enkele opmerkingen naar aanleiding van de Praeadviezen van prof. dr. Kohlbrugge en Mr. Ries over de vraag: welken invloed zal de economische ontwikkeling der laatste jaren uitoefenen op het bevolkingscijfer, in het bijzonder van Nederland? *Economisch-Statistische Berichten*, Vol. 7, pp. 852-855.

Methorst, H.W. (1923), Het jongens-overschot onder de geborenen. *De Economist*, pp. 489-501.

Methorst, H.W. (1927), Results of differential birth rate in the Netherlands. In: M. Sanger (ed.), *Proceedings of the world population conference*. London, pp. 169-189.

Methorst, H.W. (1930), XIXe zitting van het Institut International de Statistique September 1930 te Tokyo. *Economisch-Statistische Berichten* 5 November 1930, pp. 979-981.

Methorst, H.W. (1937), Leeftijdspyramide: de spil van het bevolkingsvraagstuk. *Tijdschrift voor Sociale Geneeskunde*, Augustus 1937.

Methorst, H.W. (1943), Het vijftigjarig jubileum der Centrale Commissie voor de Statistiek (1892-1942). *Economisch-Statistische Berichten*, nr. 1, pp. 2-4.

Methorst, H.W. and M.J.Sirks (1948), *Het Bevolkingsvraagstuk*. Amsterdam: Scheltema en Holkema.

Mooij, J. (1995), *Denken over welvaart. Koninklijke Vereniging voor de Staathuishoudkunde 1849-1994*. Utrecht.

NIVS (Nederlandsch Instituut voor Volkshuisvesting en Stedebouw; 1929), *Rapport over de methode-Halle voor de bepaling van de woningbehoefte*. Off-print "*Gemeentebestuur*".

NIVS (1942; 2nd ed. 1943), *Handleiding bij de voorbereiding van uitbreidingsplannen*. Alphen aan den Rijn: N. Samson N.V.

Noordman, J. (1989), *Om de kwaliteit van het nageslacht. Eugenetica in Nederland, 1900-1950*. Nijmegen: SUN.

Noordman, J. (1991/1992), Darwinisme en sociale selectie. Sociaal-darwinistische visies op evolutie en geschiedenis. *Wijsgerig perspectief*, Vol. 32, nr. 4, pp. 104-110.

Nota van Toelichting tot het Algemeen Uitbreidingsplan van Amsterdam (1935), Deel II, Bijlage IV. *Het aantal woningen, waarop in het plan gerekend moet worden*. Amsterdam, pp. 85-127.

Novotny, H. (1992), Time and social theory: towards a social theory of time. *Time and Society*, Vol. 1, nr. 3, pp. 421-454.

Oly, Joh. C. (1924), Verandering in geboorte en sterfte. Raakt Nederland overbevolkt? *Het Verzekerings-Archief*, Vol. 5, pp. 149-170.

Oly, Joh. C. (1930), Ir. F.W.'t Hooft. Het bevolkingsvraagstuk. *Het Verzekerings-Archief*, Vol. 11, pp. 163-173.

Ons Nageslacht (1929).
Paulinus, G. (1934), *Prolegomena zu einer Bevölkerungsprognose*. Inauguraldissertation. Leipzig.
Paulinus, G. (1935/36), Der innere Zusammenhang der wichtigsten bevölkerungsstatistischen Grössen und die Grenzen der Vorausberechenbarkeit der Bevölkerungsgrösse. *Allgemeines Statistisches Archiv* , Vol. 25, pp. 143-155.
Pearson, E.S. (1965/1970), Some incidents in the early history of biometry and statistics, 1890-1894. In: E.S. Pearson and M.G. Kendall (eds.; 1970), *Studies in the history of statistics and probability*. London: Griffin. Reprint from *Biometrika* , Vol. 52 (1965), pp. 3-18.
Pearson, E.S. (1967/1970), Some reflexions on continuity in the development of mathematical statistics, 1885-1920. In: E.S. Pearson and M.G. Kendall (eds.; 1970), *Studies in the history of statistics and probability*. London: Griffin. Reprint from *Biometrika* 54 (1967), pp. 341-355.
Pflaumer, P. (1988), Methoden der Bevölkerungsvorausschätzung unter besonderer Berücksichtigung der Unsicherheit. *Volkswirtschaftliche Schriften*, Heft 377. Berlin: Duncker und Humblot.
Pierson, N.G. (1876), *Grondbeginselen der staathuishoudkunde*,Tweede deel. Haarlem: Erven F. Bohn.
Pierson, N.G. (1890/1913), *Leerboek der Staathuishoudkunde*, Deel II (Revised 1913 edition by C.A. Verrijn Stuart). Haarlem: Erven F. Bohn.
Pitt-Rivers, G.I.H.F. (ed.) (1932), *Problems of proceedings of the second general assembly of the international union for the scientific investigation of population problems, London, June 15-18, 1931*. London: George Allen and Unwin Ltd.
Plasmeijer, H. (1995), Wie was Gerhard A.H. Wiebols (1895-1960)? Over een vergeten kritiek op de mainstream. *Tijdschrift voor Politieke Ekonomie*, Vol. 18, nr. 1, pp. 105--118.
Popper, K. (1957/1960), *The poverty of historicism*. Ark paperbacks, London and New York.
Pressat, R. (1983), *L'analyse démographique*. Paris: P.U.F.
Price Williams, R. (1880), On the increase of population in England and Wales, *Journal of the Statistical Society*, Vol. 43, pp. 462-496.
Pronovost, G. (1989), The sociology of time. *Current sociology. The Journal of the International Sociological Association*, Vol. 37, nr. 3, pp. 1-129.
Ptoukha, M. (1931), La Population de l'Ukraine jusque'en 1960. *Bulletin de l'Institut International de Statistique*, Tome XXV, 3ème Livraison, pp. 59-88.
Rapport van de commissie ter bestudeering van het vraagstuk van den bouw eener Tuinstad of van tuindorpen in de omgeving van Amsterdam (1929). Amsterdam: Stadsdrukkerij.
Reitsma, J.M.I. (1935), Nederland's bevolkingsaanwas. *De Ingenieur*, no. 19. Algemeen gedeelte, 19, 11-5-1934, A. 176-180.
Report by dr. Mayr on the Exhibition on Charts and diagrams at the Ninth International Statistical Congress, held at Buda Pesth, in September 1876, *Journal of the Statistical Society*, Vol. 40 (1877), pp. 131-136.
Ries, C.A. (1921), Malthus. *De Economist* pp. 549-561; pp. 602-619.
Rikkert, A.J.A. (1919), De berekening van de woningbehoefte. *Gemeentebelangen*, Vol. 14, nr. 15, pp. 225-228; nr.16, pp. 244-246.

Robbins, L. (1927), The optimum theory of population. In: T.E. Gregory and H. Dalton (eds.), *London essays in economics: In honour of Edwin Cannan*. London: George Routledge and Sons Ltd, pp. 103-134.

Roegholt, M.J.W. (1925), *Het Stadsgewest. Een studie over den trek naar de steden en zijn gemeentelijk-economische gevolgen*. PhD thesis Techn. Univ. Delft. Wassenaar: N.V. Uitgevers Maatschappij v/h G. Delwel.

Rogers, A. (1985), *Regional population projection models*. Beverly hills, London. New Delhi: Sage Publications.

Rooy, M. (1920), Bevolkingstoeneming en woningnood. *Economisch-Statistische Berichten*, Vol. 5, pp. 1108-1110.

Rooy, M. (1921), De woningbehoefte in de eerste twintig jaren. *Economisch-Statistische Berichten*, Vol. 6, pp. 656-658.

Röpke, W. (1930), Sozialökonomische Betrachtungen über den abnehmenden Bevölkerungszuwachs. *De Economist*, Vol. 79, pp. 637-655.

Royston, E. (1956/1970), A note on the history of the graphical presentation of data. In: E.S. Pearson and M.G. Kendall (eds.; 1970), *Studies in the history of statistics and probability*, London: Griffin. Reprint from *Biometrika*, Vol. 43 (1956), pp. 241-247.

Rückert, J.H.E. (1913), Stedenbouw, een arbeidsveld van den ingenieur bedreigd. *De Ingenieur*, Vol. 28, nr. 48, pp. 1034-1036.

Rückert, J.H.E. (1914), Stedenbouw. *De Ingenieur*, Vol. 29, nr. 10, pp. 199-200.

Rückert, J.H.E. (1917), *Rapport van den heer J.H.E.Rückert, Directeur van Publieke Werken, met betrekking tot het Algemeen Uitbreidingsplan der gemeente Tilburg*. Tilburg.

Ryder, N.B. (1965), The cohort as a concept in the study of social change. *American Sociological Review*, Vol. 30, pp. 843-861.

Saltet, R.H. and Ph. Falkenburg (1907), Kindersterfte in Nederland (in de jaren 1881-1905). *Statistische Mededeelingen* uitgegeven door het Bureau van Statistiek der Gemeente Amsterdam, no. 19. Amsterdam: Johannes Müller.

Samuelson, P.A. (1977), Resolving a historical confusion in population analysis. In: D. Smith and N. Keyfitz, *Mathematical demography. Selected papers*. Berlin: Springer Verlag, 109-129.

Sanger, M. (ed.) (1927), *Proceedings of the world population conference*. Geneva, August 29th to September 3rd, 1927. London: Edward Arnold and Co.

Sauvy, A. (1928), La population française jusqu'en 1956. Essai de prévision démographique. *Journal de la Société de Satistique de Paris* 70, 12, pp. 321-329.

Sauvy, A. (1932), Calculs démographiques sur la population française jusqu'en 1980. *Journal de la Société de Statistique de Paris*, pp. 338-347. Discussion: pp. 338-347.

Schutte, W.F.M. (1922), Berekening van de woningbehoefte in Utrecht. *Gemeentebelangen*, Vol. 1, nr. 27, pp. 214-216.

Séances des sections. *Bulletin de l'Institut International de Statistique*, Tome 25, Iére Livraison, 1931. Tokyo, pp. 66-74.

Shryock, H.S and J.S.Siegel (1973), *The methods and materials of demography., Vol. 2*. Washington: U.S. Bureau of the Census.

Smit, F. (undated), *Van Amsterdamse Huize. Ontwikkeling en identiteit van Het Woningbedrijf Amsterdam*. Amsterdam: Stichting Het Woningbedrijf Amsterdam.

Smith, D. and N. Keyfitz (1977), *Mathematical demography. Selected papers*, Berlin, Heidelberg, New York. Springer Verlag.

Soloway, R.A. (1990), *Demography and degeneration. Eugenics and the declining birth rate in Twentieth-Century Britain*. Chapel Hill and London: The North Carolina University Press.
Spengler, J.J. (1936), Population prediction in nineteenth century America. *American Sociological Review*, Vol. 1, pp. 905-921.
Spengler, J.J. (1938/1968), *France faces depopulation*. New York: Greenwood Press (repr. 1968).
Spier, J. (1930), Zal de bevolking blijven aangroeien? Boekbespreking Ir. F.W. 't Hooft. Het bevolkingsvraagstuk. Waar wij heengaan in West-Europa? H.J. Paris, Amsterdam 1929. *De Socialistische Gids*, uitgeverij N.V. De Arbeiderspers, Amsterdam, juli 1930.
Stamhuis, Ida H. (1989), *'Cyfers en aequaties' en 'Kennis der staatskrachten'. Statistiek in Nederland in de negentiende eeuw*. PhD thesis, V.U. Amsterdam: Rodopi.
Statistisches Reichsamt (1928), Ausblick auf die zukünftige Bevökerungsentwicklung im Deutschen Reich. In: *Statistiek des Deutschen Reichs*, Band 401,I. Volkszählung. Die Bevölkerung des Deutschen Reichs nach den Ergebnissen der Volkszählung 1925. Berlin. Teil 1, pp. 641-680.
Stolzenburg, R. (1984), *Het sociaal-wetenschappelijk onderzoek bij het uitbreidings- en structuurplan. Een sociografische opgave*. Eindhoven.
Streekplan Eindhoven en omgeving, Bijl.II "Bevolking", mei 1942.
Stübben, J. (1890), *Der Städtebau. Handbuch der Architektur*. Vierter Teil, 9. Halb-Band. Darmstadt: Verlag von Arnold Bergstarsser. Reprint der 1. Auflage, Braunschweig/-Wiesbaden: Friedr. Vieweg und Sohn.
Szreter. S. (1993), The idea of demographic transition and the study of fertility change: A critical intellectual history. *Population and Development Review*, Vol. 19, nr. 3, pp. 659-701.
Teitelbaum, M.S. and J.M. Winter (1985), *The fear of population decline*, Orlando: Academic Press, Inc.
Ter Heide, H. (1973), Uitgangspunten en begrippenapparaat voor regionale bevolkingsvoor-uitberekeningen. *Bevolking en Gezin*, nr.3, pp. 445-470.
Ter Heide, H. (1992), Bevolking en ruimte; verschuivend perspectief in onderzoek en beleid. In: A. Kuijsten and A. Assen (eds., 1992), *BeVOLking en RUIMte*. Inleidingen gehouden op de jaarvergadering van de Nederlandse Vereniging voor Demografie te Voorburg op 11 oktober 1991. NvD, pp. 83-104.
Ter Heide, H. (1998 forthcoming), Een halve eeuw regionale bevolkingsvooruitberekeningen in Nederland. *Bevolking en Gezin*.
Thijsse, Jac. P. (1963), *Enkele gedachten over het ruimtelijk beeld van Nederland in het jaar 2000 bij een bevolking van 20 miljoen*. Publikatie nr. 6. The Hague: Vereniging voor Demografie.
Thompson, W.S. (1931A), Probable future growth of population in the United States. *Bulletin de l'Institut International de Statistique*, 3éme Livraison. Tokyo, 1931, pp. 89-97.
Thompson, W.S. (1931B), *Ratio of children to women 1920*. Washington: United States Government Printing Office.
't Hooft, F.W. (1926), De maximum-bevolking van Nederland. *De Economist*, Vol. 75, pp. 690-707.
't Hooft, F.W. (1927a), De maximum-bevolking van Nederland. De theorie van den transporteur. *De Economist*, Vol. 76, pp. 55-61.

't Hooft, F.W. (1927b), De bevolkingsvermeerdering in Nederland. *De Economist*, Vol. 76, pp. 659-676.

't Hooft, F.W. (1927c), Naschrift. *De Economist*, Vol. 76, pp. 318-319.

't Hooft, F.W. (1928), Sterftetafels, sterftecijfers en generatietafels. *De Economist*, Vol. 77, pp. 511-535.

't Hooft, F.W. (1929), *Het bevolkingsvraagstuk. Waar wij heengaan in West-Europa.* Amsterdam: H.J. Paris.

't Hooft, F.W. (1931), Over den aanwas van bevolkingen – Duitschland en Nederland. *De Economist*, Vol. 80, pp. 267-281.

't Hooft, F.W. (1936), De bevolking van Nederland in de crisis. *De Economist*, Vol. 85, pp. 99-111.

't Hooft, F.W. (1937), Over het verstandig gebruik van formules en methoden. *De Economist*, Vol. 86, pp. 370-375

Ubbink, J.B. (1937), Foutieve formules in het bevolkingsvraagstuk. *De Economist*, Vol. 86, pp. 195 and 654-655.

United Nations (1958), *Multilingual demographic dictionary. English Section*. New York.

United Nations (1973; 1978), *The determinants and consequences of population trends. New summary of findings on interaction of demographic, economic and social factors*, Vol. I (1973); Vol. II (1978) Bibliography and Index. New York.

Van Beusekom, H.G. (1952), Organen bij de uitvoering van de wet. A. Het Rijk. In: *50 jaar Woningwet 1902-1952*. Alphen aan den Rijn: N. Samson N.V., pp. 25-48.

Van Braam, A. (1948), Over de methode in de survey ten behoeve van een stedelijk uitbreidingsplan -in- hoofdzaak. *Tijdschrift voor Volkshuisvesting en Stedebouw* 29, pp. 165-169.

Van de Kaa, D.J. (1987), Europe's second demographic transition. *Population Bulletin* (Population Reference Bureau, Inc.), 42, 1.

Van de Kaa, D. J. (1996), Anchored narratives: The story and findings of half a century of research into the determinants of fertility. *Population Studies*, Vol. 50, pp. 389-432.

Van den Brandhof, M. (undated), *J.H.E. Rückert (1879-1918)*. Unpublished typoscript.

Van den Brekel, J.C. (1996), De CBS-bevolkingsprognose van 1965; een 'blooper' van jewelste? *Bevolking en Gezin* 2, pp. 123-142.

Van der Flier, G., S.O. van Poelje and H. van der Weijde (eds.; 1952), *50 jaar Woningwet 1902-1952. Gedenkboek*. Alphen aan den Rijn: N. Samson N.V.

Van der Heiden, N. (1990), *Bouwen met beleid. De doorwerking van het provinciaal ruimtelijk beleid in Noord-Holland 1926-1943*. Werkstukken, nr. 116. Amsterdam: Planologisch en Demografisch Instituut.

Van der Kaa, H. (1925), De methode van Halle. *Tijdschrift voor Volkshuisvesting en Stedebouw* 3, pp. 64-70.

Van der Schaar, J. (1980), Volkshuisvesting. Ontwikkelingen in beleid en denkbeelden tot de Tweede Wereldoorlog. *Tijdschrift voor Stedebouw en Volkshuisvesting* pp. 1-78.

Van der Schaar. J. (1991), *Volkshuisvesting: een zaak van beleid*. Aula paperback 198. Utrecht: Het Spectrum b.v.

Van der Valk, A. (1989), *Amsterdam in aanleg. Planvorming en handelen 1850-1900*. Planologische studies 8. Amsterdam: Planologisch en Demografisch Instituut.

Van der Valk, A. (1990), *Het levenswerk van Th.K. van Lohuizen 1890-1956. De eenheid van het stedebouwkundig werk*. Delft: Delft University Press.

Vandeschrick, C. (1992), Le diagramme de Lexis revisité. *Population* 47, nr. 5, pp. 1241-1262.
Vandeschrick, C. (1994), Le temps dans le temps en démographie. Le diagramme de Lexis: bilan et perspectives. In: Éric Vilquin (ed.) (1994), *Le temps et la démographie*. Chaire Quételet 1993. Louvain-la-Neuve: Academia/L'Harmattan, pp. 271-308.
Van Doorn, J.W.M. and F.A. van Vught (eds.; 1981), *Nederland op zoek naar zijn toekomst*. Aula-paperback 65. Utrecht: Het Spectrum.
Van Fulpen, H. (1985), *Volkshuisvesting in demografisch en economisch perspectief*. Sociaal en Cultureel Planbureau. Sociale en Culturele Studies – 8. 's-Gravenhage: Staatsuitgeverij.
Van Haaften, M. (1923), De sterftetafels, afgeleid uit de volkstellingen in Nederland. *Het Verzekeringsarchief*, Vol. 4, pp. 128-163.
Van Imhoff, E., A. Kuijsten, P. Hooimeijer and L. van Wissen (eds.)(1995), *Household demography and household modeling*. New York and London: Plenum Press.
Van Kesteren, G. (1984), *Streekplanning in Noord-Holland 1902-1943*. Verkenningen, nr. 34. Amsterdam: Planologisch en Demografisch Instituut.
Van Lohuizen, Th.K. (1922), *Zwei Jahre Wohnungstatistik in Rotterdam. Eine neue Methode der Statistik über Wohnungsbedarf und Wohnungsvorrat*. Herausgegeben vom Nederlandsch Instituut voor Volkshuisvesting. Berlin: Carl Heymanns Verlag.
Van Lohuizen, Th.K. (1925), Concentratie en decentralisatie; de bevolkingsbeweging in de stedelijke invloedssfeer Holland-Utrecht 1869-1920. *Tijdschrift voor Economische Geografie*, nr. 16, pp. 341-350.
Van Lohuizen, Th.K. (1948), De eenheid van het stedebouwkundig werk. *Tijdschrift voor Economische en Sociale Geografie*, Vol. 39, nr. 5, pp. 401-108.
Van Lohuizen, Th.K. and G.Th.J. Delfgaauw (1935), De toekomstige verdeeling van bevolking en van woningvoorraad over het land. *Tijdschrift voor Volkshuisvesting en Stedebouw*, Vol. 16, nr. 11, pp. 203-209; nr. 12, pp. 226-230.
Van Maarseveen, J.G.S.J. (ed.) (1992/1993), *Briefwisseling van Nicolaas Gerard Pierson 1839-1909*. Deel II (1885-1897); Deel III (1898-1909). Amsterdam: De Nederlandsche Bank, N.V.
Van Pesch, A.J. (1912), *Sterftetafels voor Nederland, afgeleid uit de periode 1900-1909*. Aanhangsel tot het 2e Deel van de Uitkomsten der Negende tienjaarlijkse Volkstelling in het Koninkrijk der Nederlanden gehouden op 31 december 1990. 's-Gravenhage.
Van Poppel, F. (1992), *Trouwen in Nederland. Een historisch-demografische studie van de 19e en vroeg-20e eeuw*. A.A.G. Bijdragen 33. Wageningen: Landbouwuniversiteit.
Van Praag, Ph. (1976), *Het bevolkingsvraagstuk in Nederland. Ontwikkeling van standpunten en opvattingen (1918-1940)*. NIDI-publikaties 1. Deventer: Van Loghum Slaterus.
Van Praag, Ph. (1977), Views and concepts relating to population problems in the Netherlands 1918-1939. *Population Studies*, Vol. 31, nr. 2, pp. 251-265.
Van Praag, Ph. (1980), Prof. L.H.J. Angenot overleden. Demografische Kroniek. *Bevolking en Gezin*, nr. 1, p. 145.
Van Rossem, V. (1988), Berlage: beschouwingen over stedebouw 1892-1914. In S.Polano, *Hendrik Petrus Berlage. Het complete werk*. Alphen aan de Rijn: Atrium, pp 46-66.
Van Sandick, R.A. (1921), *De Ingenieur*, Vol. 36, p. 519.
Van Seventer, H. and A.P. Vossen (1979), *Enkele theoretische en methodologische grondslagen van de bevolkingsprognostiek: een eerste verkenning*. Tilburg: Katholieke Hogeschool.

Van Vuuren, L. (1936), *Een waarschuwend woord aan het Nederlandsche volk*. Naar aanleiding van: W.A. Bonger – "De stand van het bevolkingsvraagstuk in Nederland". Utrecht: N.V. A. Oosthoek's Uitgevers Maatschappij.

Van Vuuren, L. (1937), Foutieve formules in het bevolkingsvraagstuk. *De Economist*, Vol. 86, pp. 375-377.

Van Zanten, J.H. (1925), De methode van Halle. *Tijdschrift voor Volkshuisvesting en Stedebouw* , Vol. 6, pp. 145-147.

Van Zanten, J.H. (1926), G.A.H. Wiebols. De toekomstige bevolkingsgrootte van Nederland. Boekbespreking. *Mens en Maatschappij*, pp. 344-345.

Van Zanten, J.H. (1927), *Leerboek der statistische methode*. Alphen aan den Rijn: N. Samson N.V.

Van Zanten, J.H. (1930), Boekbespreking van Ir. F.W.'t Hooft, Het bevolkingsvraagstuk. Waar wij heengaan in West-Europa. Amsterdam: H.J. Paris, 1929, 111 blz. *De Economist*, Vol. 79, nr. 4, pp. 318-320.

Van Zanten, J.H. (1931), Boekbespreking Ir. F.W. 't Hooft Het bevolkingsvraagstuk. Waar wij heengaan in West-Europa (1929). *Mens en Maatschappij* 7, pp. 64-67.

Van Zanten, J.H. (1933a), Ir. P. Bakker Schut, De bevolkingsbeweging in Nederland en in het bijzonder te 's-Gravenhage. *De Economist*, Vol. 82, pp. 589-591.

Van Zanten, J.H. (1933b), Boekbespreking De toekomstige bevolking van Amsterdam. Studie betreffende den toekomstigen aanwas van Amsterdam, uit de serie rapporten van den Dienst der Publieke Werken, dienende als 'Grondslagen voor de stedebouwkundige ontwikkeling van Amsterdam' 1932. *Mensch en Maatschappij*, Vol. 9, pp. 457-459.

Van Zanten, J.H. (1938), *Leerboek der statistische methode*. Alphen aan den Rijn: N. Samson N.V., derde druk.

Verrijn Stuart, C.A. (1905a), Verslag van de tiende zitting van het Institut International de Statistique (London 31 Juli - 5 Augustus 1905). No. 101. Bijvoegsel tot de *Nederlandsche Staatscourant* van Zondag 22 en Maandag 23 October 1905, No.248. 14 pp.

Verrijn Stuart, C.A. (1905b), De tiende zitting van het Institut International de Statistique. De Economist , Vol. 49, pp. 733-748.

Verrijn Stuart, C.A. (1910), *Inleiding tot de beoefening der statistiek*. Eerste deel. De statistische methode en hare toepassing op het gebied der demografie. Haarlem: Erven F.Bohn.

Verrijn Stuart, C.A. (1919), De daling van het geboortecijfer in Nederland. *Economisch-Statistische Berichten*, Vol. 4, pp. 79 82.

Verrijn Stuart, C.A. (1920), *De grondslagen der volkshuishouding*. Haarlem: De erven F.Bohn.

Verrijn Stuart, C.A. (1921), De voorloopige uitkomsten der volkstelling. *Economisch-Statistische Berichten*, Vol. 6, pp. 467-469.

Verrijn Stuart, C.A. (1922), De toeneming der bevolking in Nederland. *Economisch-Statistische Berichten*, pp. 900-902.

Verrijn Stuart, C.A. (1924), *De grondslagen der volkshuishouding*. Second (revised) printing, Haarlem: De erven F. Bohn.

Verrijn Stuart, C.A. (1925), Boekbespreking van Dr. G.A.B. Wiebols. De toekomstige bevolkingsgrootte in Nederland. Rotterdam 1925. *De Economist*, Vol. 74, pp. 447-453.

Verrijn Stuart, C.A. (1928; 2nd revised ed. of 1910), *Inleiding tot de beoefening der statistiek*. Eerste deel. Haarlem: De erven F. Bohn.

Verrijn Stuart, C.A. and H.W. Methorst (1907), Verslag van het verhandelde op de elfde bijeenkomst van het Internationaal Statistisch Instituut, gehouden te Kopenhagen van 26 tot 31 Augustus 1907. Bijvoegsel tot de *Nederlandsche Staatscourant*, Vrijdag 6 December 1907, no. 286, pp. 1-18.

Verslag van de Algemeene Vergadering gehouden te Den Haag (1922). Vereeniging voor de Staathuishoudkunde en de Statistiek. Den Haag: Martinus Nijhoff.

Verwey, A.J. (1875), Principles of vital statistics. *Journal of the Statistical Society*, Vol. 28, pp. 487-513.

Vilquin, É. (ed.) (1994), *Le temps et la démographie. Chaire Quetelet 1993*. Louvain-la-Neuve: Academia/L'Harmattan.

Von Mises, R. (1932), Altersschichtung und Bevölkerungszahl in Deutschland. *Die Naturwisschenschaften*, Vol. 20, nr. 4, pp. 59-62.

Von Mises, R. (1933), Über die Vorausberechnung vom Umfang und Altersschichtung der Bevölkerung Deutschlands. *Blätter für Versicherungs-Mathematik*, Vol. 2, nr. 10, pp. 359-371.

VSS (Vereeniging van Staathuishoudkunde en de Statistiek, 1908) *Welke gedragslijn moet aan gemeenten worden aanbevolen in het belang van hare doelmatige uitbreiding? Zijn er wetswijzigingen noodig om deze gedragslijn mogelijk te maken? Zoo ja, welke?* Pre-adviseurs: J.H. Valckenier Kips, H.J. Nieboer, J.W.C. Tellegen. The Hague: Martinus Nijhoff.

VSS (1922), *Welken invloed zal de economische ontwikkeling der laatste jaren uitoefenen op het bevolkingscijfer, in het bijzonder van Nederland?* Pre-adviseurs: J.H.F. Kohlbrugge and L.A.Ries. The Hague: Mart. Nijhoff, pp. 1-108.

Vulhopp, F. (1932), La population belge de demain. Étude démographique. *Annales de la Société Scientifique de Bruxelles* tome 52, pp. 28-46.

Wagenaar, C. (1992), *Welvaartsstad in wording. De wederopbouw van Rotterdam 1940-1952*. Rotterdam: NAi.

Wattelar, Chr. (1980), Les perspectives démographiques par sexe et par age: simplisme ou sophistication? In: *Perspectives de population, d'emploi et de croissance urbaine*. Chaire Quetelet '80. Louvain-la-Neuve (Belgium): Ordina Editions, pp. 57-88.

Weindling, P. (1989), Fascism and population in comparative European perspective. In: M.S. Teitelbaum and J.M. Winter (eds), Population and resources in Western intellectual traditions. *Population and Development Review*. Supplement to Volume 14. Cambridge, New York: Cambridge University Press.

Weindling, P. (1990), *Health, race and German politics between national unification and Nazism, 1870-1945*. Cambridge, New York: Cambridge University Press.

Westergaard, H. (1908), The Horoscope of the population in the Twentieth Century, *Bulletin de l'Institut International de Statistique* Tome XVII. Première Partie. Compte-Rendue de la XIe Session de l'Institut International de Statistique à Copenhague du 26 Août 1907. Copenhague, pp. 103-117.

Westergaard, H. (1932), *Contributions to the history of statistics*. London: P.S. King and Sons.

Whelpton, P.K. (1936), An empirical method of calculating future populations. *Journal of the American Statistical Association*, Vol. 31, pp. 457-473.

Wiebols, G.A.H. (1925), *De toekomstige bevolkingsgrootte in Nederland*. PhD thesis, Handelshogeschool Rotterdam. Vlaardingen.

Wiebols, G.A.H. (1927), De maximum-bevolking van Nederland. *De Economist*, Vol. 76, pp. 315-318.
Wiebols, G.A.H. (1933), *Waarde en prijs*. Amsterdam.
Willekens, F.J. (1990), Demographic forecasting: State of the art and research needs. In: C.A. Hazeu and G.A.B. Frinking (eds.), *Emerging issues in demographic research*, Amsterdam: Elsevier, pp. 9-66.
Willekens, F. (1992), Models of man in demography. In: H.A. Becker (ed.), *Dynamics of cohort and generations research*. Amsterdam: Thesis Publishers, pp. 253-281.
Willekes MacDonald, P.J. (1933), De bevolkingsafname van Amsterdam. *De Ingenieur* A, pp. 206-209.
Wolfe, A.B. (1928/1929), The population problem since the World War: A survey of literature and research. *The Journal of Political Economy*, Vol. 36 (1928), pp. 529-559 and pp. 662-685; Vol. 37 (1929), pp. 87-120.
Yntema, L. (1977), *Inleiding tot de demometrie*. NIDI-publikaties 3.Deventer: Van Loghum Slaterus.
Yule, G. Udny (1925), The growth of population and the factors which control it. *Journal of the Royal Statistical Society*, Vol. 88, pp. 1-58. Proceedings of the meeting, pp. 58-62.
Zahn, Fr. (1934), *50 Années de l'Institut International de Statistique*.
Zonneveld, W. (1991), *Conceptvorming in de ruimtelijke planning. Patronen en processen*. Planologische Studies 9a. Amsterdam: Planologisch en Demografisch Instituut.

Archives[1]

Archief LHJ Angenot. Nederlands Architectuurinstituut (NAi), Rotterdam

Archief Centrale Commissie voor de Statistiek (CCS), Voorburg

Archief De Economist, Groningen

Archief Faculteit Economische Wetenschappen Universiteit van Amsterdam, Amsterdam

Archief International Statistical Institute (ISI), Voorburg

Archief Th.K. van Lohuizen. Nederlands Architectuurinstituut (NAi), Rotterdam

Archief Nederlands Centraal Bureau voor de Statistiek (NCBS), Voorburg

Archief Nederlandsch Comité van de Internationale Vereniging voor de Wetenschappelijke Bestudering van de Bevolking, NIDI, The Hague

Archief Rijksdienst voor het Nationale Plan (Arch. RNP), Ministerie VROM, The Hague.

Archief Stichting voor het bevolkingsonderzoek in de drooggelegde Zuiderzee-polders. Sociaal Historisch Centrum voor Flevoland, Lelystad

Edwin Cannan archives. London School of Economy and Political Sciences, London, United Kingdom

---

1 All archives in the Netherlands unless stated otherwise.

Gemeente Archief Amsterdam (GAA), Amsterdam

Rijksarchief in Noord-Holland (RANH), Haarlem

Rijksinstituut voor Oorlogsdocumentatie (RIOD), Amsterdam

Index of Items

Index of Names

NIDI CBGS Publications

1. *Population and family in the Low Countries I* (1976). 179 pp.
2. R.L. and R. Schoenmaeckers (1976), *From incidental to planned parenthood. Results of the second nationional fertility survey in Belgium.* 145 pp.
3. R. Chester (ed.) (1977), *Divorce in Europe.* 316 pp.
4. M. Leroy (1978), *Population and world politics.* 144 pp.
5. G. Santow (1978), *A simulation approach to the study of human fertility.* 215 pp.
6. *Population and family in the Low Countries II* (1978). 153 pp.
7. M. Niphuis-Nell (ed.) (1978), *Demographic aspects of the changing status of women in Europe.* 153 pp.
8. G. Dooghe and J. Helander (eds.) (1979), *Family life in old age.* 203 pp.
9. G.E. Wiersma (1983), *Cohabitation, an alternative to marriage?: a cross-national study.* 204 pp.
10. *Population and family in the Low Countries III* (1984). 201 pp.
11. R.L. Cliquet and J. Balcaen (1983), *Quantitative and qualitative aspects of family-size variation.* 192 pp.
12. *Population and family in the Low Countries IV (1987).* 151 pp.
13. F.H.A.G. Zwart (1986), *Population growth of Fiji.* 248 pp.
14. A.C. Kuijsten (1986), *Advances in family demography.* 383 pp.
15. F. Deven and R.L. Cliquet (1986), *One parent families in Europe.* 365 pp.
16. *Population and family in the Low Countries V* (1987). 151 pp.
17. H. Moors and J. Schoorl (eds.) 1989, *Lifestyles , contraception and parenthood.* 407 pp.
18. R.L. Cliquet et al (eds.) (1989), *Population and family in the Low Countries VI.* 156 pp.
19. R.L. Cliquet and L. Vanden Boer (eds.) (1989), *Economic and social implications of aging in the ECE region.* 225 pp.
20. N.W. Keilman (1990), *Uncertainty in national population forecasting: Issues, backgrounds, analyses, recommendations.* 211 pp.
21. R.L. Cliquet (ed.) (1991), *Desirabilities and possibilities of a fertility recovery at replacement level in Europe.* 141 pp.
22. G.C.N. Beets et al. (eds.) (1991), *Population and Family in the Low Countries 1991.* 217 pp.
23. E. van Imhoff and N. Keilman (1991), *LIPRO 2.0: and application of a dynamic demographic projection model to household structure in the Netherland.* 245 pp.
24. N. Keilman and H. Cruijsen (eds.) (1992), *National population forecasting in industrialized countries.* 364 pp.

25. E. van Imhoff, E. Themmen and F. Willekens, (eds.) (1992), *Population, environment, and development*. 89 pp.
26. G.C.N. Beets, R.L. Cliquet, G. Dooghe and J. de Jong Gierveld (eds.) (1993), *Population and Family in the Low Countries 1992: Family and labour*. 270 pp.
27. N. van Nimwegen, J.C. Chesnais and P.A. Dykstra (1993), *Coping with sustained low fertility in France and the Netherlands*. 321 pp.
28. R. Penninx, J. Schoorl and C. van Praag (1993), *The impact of international migration on receiving countries: the case of the Netherlands*. 251 pp.
29. G. Dooghe and L. Vanden Boer (eds.) (1993), *Sheltered accomodation for elderly people in an international perspective*. 148 pp.
30. G.C.N. Beets, J.C. van den Brekel, R.L. Cliquet, G. Dooghe and J. de Jong Gierveld (eds.) (1993) *Population and Family in the Low Countries 1993: late fertility and other current issues*. 246 pp.
31. E. van Imhoff, L. van Wissen and K. Spieβ (1994), *Regional population projections in the countries of the European Economic Area*. 145 pp.
32. M. Corijn (1996), *Transition into adulthood in Flanders. Results from the fertility and family survey 1991-92*. Brussel: Ministerie van de Vlaamse Gemeenschap, departement Leefmilieu en Infrastructuur, afdeling Logistiek - sectie Drukkerij, 216 pp.
33. J.C. van den Brekel and F. Deven (eds.) (1997), *Population and family in the Low Countries 1996/1997: Selected current issues*. Brussel: Ministerie van de Vlaamse Gemeenschap, departement Leefmilieu en Infrastructuur, afdeling Logistiek - sectie Drukkerij, 244 pp.

- Number 1-19 and 32 and further can be obtained, if still available, via the NIDI, numbers 20-31 only in bookshops or directly from the publisher (Swets & Zeitlinger, Publishing Division, P.O. Box 825, 2160 SZ Lisse, The Netherlands; telephone:(int +31)(0)252 435 111; telefax: (int + 31)(0)252 415888

European Studies of Population

1. J.-P. Gonnot, N. Keilman and C. Prinz: *Social Security, Household, and Family Dynamics in Ageing Societies*. 1995 ISBN 0-7923-3395-0
2. H. van den Brekel and F. Deven (eds.): *Population and Family in the Low Countries 1994*. Selected Current Issues. 1995 ISBN 0-7923-3396-9
3. R. Cliquet and C. Thienpont: *Population and Development*. A Message from the Cairo Conference. 1995 ISBN 0-7923-3763-8
4. H. van den Brekel and F. Deven (eds.): *Population and Family in the Low Countries 1995*. Selected Current Issues. 1996 ISBN 0-7923-3945-2
5. H.A. de Gans: *Population Forecasting 1895–1945*. The Transition to Modernity. 1999 ISBN 0-7923-5537-7

KLUWER ACADEMIC PUBLISHERS – DORDRECHT / BOSTON / LONDON

GPSR Compliance

The European Union's (EU) General Product Safety Regulation (GPSR) is a set of rules that requires consumer products to be safe and our obligations to ensure this.

If you have any concerns about our products, you can contact us on ProductSafety@springernature.com

In case Publisher is established outside the EU, the EU authorized representative is:

Springer Nature Customer Service Center GmbH
Europaplatz 3
69115 Heidelberg, Germany

Batch number: 10755284

Printed by Printforce, the Netherlands